Biochemistry of Virus-Infected Plants

RESEARCH STUDIES IN BOTANY AND RELATED APPLIED FIELDS

Series Editor: **Dr. P. S. Nutman, F.R.S.**

1. Root Nodules of Legumes: Structure and Functions
   F. J. Bergersen

2. Acid Soil and Acid Rain
   The Impact on the Environment of Nitrogen and Sulphur Cycling
   I. R. Kennedy

3. Biochemistry of Virus-Infected Plants
   R. S. S. Fraser

Biochemistry of Virus-Infected Plants

R. S. S. Fraser, PhD DSc

Biochemistry Section
Institute of Horticultural Research
(Formerly National Vegetable Research Station)
Wellesbourne, Warwick, England

RESEARCH STUDIES PRESS LTD.
Letchworth, Hertfordshire, England
JOHN WILEY & SONS INC.
New York · Chichester · Toronto · Brisbane · Singapore

RESEARCH STUDIES PRESS LTD.
58B Station Road, Letchworth, Herts. SG6 3BE, England

Marketing and Distribution
Australia, New Zealand, South-east Asia:
Jacaranda-Wiley Ltd., Jacaranda Press
JOHN WILEY & SONS INC.
GPO Box 859, Brisbane, Queensland 4001, Australia

Canada:
JOHN WILEY & SONS CANADA LIMITED
22 Worcester Road, Rexdale, Ontario, Canada

Europe, Africa:
JOHN WILEY & SONS LIMITED
Baffins Lane, Chichester, West Sussex, England

North and South America and the rest of the world:
JOHN WILEY & SONS INC.
605 Third Avenue, New York, NY 10158, USA

Library of Congress Cataloging-in-Publication Data

Fraser, R. S. S.
Biochemistry of virus-infected plants.
(Research studies in botany and related applied fields; 3)
Bibliography: p.
Includes index.
1. Plant viruses—Host plants. 2. Virus diseases of plants. 3. Plant immunochemistry. 4. Botanical chemistry. I. Title. II. Series.
SB736.F73 1987 632'.8 86–24794
ISBN 0 86380 046 7
ISBN 0 471 91299 9 (Wiley)

British Library Cataloguing in Publication Data

Fraser, R. S. S.
Biochemistry of virus-infected plants.—
(Research studies in botany and related applied fields; 3)
1. Virus diseases of plants
I. Title II. Series
581.2'34 SB736
ISBN 0 86380 046 7

ISBN 0 86380 046 7 (Research Studies Press Ltd.)
ISBN 0 471 91299 9 (John Wiley & Sons Inc.)

Printed in Great Britain by Galliard (Printers) Ltd, Great Yarmouth

Preface

My first involvement in research on plant viruses was in 1969. I was working in Georg Melchers' department of the Max-Planck Institut für Biologie in Tübingen, with the (then) new technique of fractionating nucleic acids on polyacrylamide gels. I remember my surprise, on first running a sample of nucleic acids from tobacco leaf infected with tobacco mosaic virus, at the sheer size of the viral RNA peak. It represented about 75% of the total nucleic acid of the leaf. Clearly there was nothing subdued about this form of pathogenesis: the plant had been turned into a virus factory. This posed several interesting questions about the interactions between virus multiplication, and host nucleic acid and protein metabolism, to which I devoted the remainder of my time in Germany. The results showed that despite the whole-hearted nature of the pathogenesis, the effects on host metabolism were generally rather subtle, and could be interpreted as being of benefit to virus multiplication in the longer term as well as in the immediate sense.

The following six years were spent going round and round the yeast cell cycle. Viruses occasionally appeared on my gels, in the form of the double-stranded RNAs of mycoviruses, but only as casual visitors, and not as direct objects of research.

In 1977, on coming to work at the National Vegetable Research Station, I resumed work on plant viruses. Again, this has mainly been concerned with the host response, and in particular the mechanisms of resistance, and biochemical controls of host growth and symptom formation.

In 1969, the amino acid sequences of the coat proteins of several viruses or strains had been determined, but very little progress had been made in sequencing any viral nucleic acid. The details of assembly of tobacco mosaic virus were beginning to be worked out, and there was basic X-ray information on particle structure. Since then, new methods have produced an explosion of information on nucleic acid and protein sequences, three-dimensional structure and assembly. The most-studied viruses are now highly defined in molecular or even in atomic terms.

The same cannot be said of the plant, or of the interaction between plant and virus. This is hardly surprising. Viruses contain perhaps a dozen or fewer genes, whereas plants may contain up to 100,000. These genes, or their functional products, need not all be involved in the host-virus interaction. But the evidence suggests that many aspects of host metabolism are altered, directly or indirectly, by infection. Some of the research activity in plant virology is now swinging from purely molecular studies of purified viruses, towards studies of the host plant. It seemed timely, therefore, to review the diverse changes in plant metabolism caused by virus infections.

It is a pleasure to thank all those who have been involved in the genesis and writing of this book. Firstly, those colleagues who have worked with me over the years on virus-related problems; much of their work is cited in the text. I am grateful to Hilary Haigh, David Walkey and Roger Wood for reading the text and for their comments, and to the Series Editor, Phillip Nutman, for many useful suggestions. One of my aims was to provide a comprehensive source of references to the primary sources, which are widely scattered; I thank Julie Payne for her efforts in maintaining my publications database. Finally, I am grateful to John Bleasdale, Director of the National Vegetable Research Station, for encouragement and for library and computer facilities.

R S S Fraser
Stratford upon Avon
July 1986

Contents

CHAPTER 6. THE BIOCHEMISTRY OF RESISTANCE TO VIRUSES

CHAPTER 7. HOW DO VIRUSES CONTROL PLANT GROWTH AND DEVELOPMENT, AND CAUSE SYMPTOMS?

Abbreviations of Virus and Viroid Names

AMV	Alfalfa mosaic virus
ASBV	Avocado sunblotch viroid
BBMV	Broad bean mottle virus
BCMV	Bean common mosaic virus
BGMV	Bean golden mosaic virus
BMV	Brome mosaic virus
BNYV	Broccoli necrotic yellows virus
BPMV	Bean pod mottle virus
BSMV	Barley stripe mosaic virus
BWYV	Beet western yellows virus
BYDV	Barley yellow dwarf virus
BYV	Beet yellows virus
CaMV	Cauliflower mosaic virus
CCCV	Coconut cadang-cadang viroid
CCMV	Cowpea chlorotic mottle virus
CEV	Citrus exocortis viroid
CLRV	Cherry leaf roll virus
CMV	Cucumber mosaic virus
CPMV	Cowpea mosaic virus
CSSV	Cocoa swollen shoot virus
CTV	Citrus tristeza virus
CuTV	Curly top virus
CYMV	Clover yellow mosaic virus
DEMV	Dolichos enation mosaic virus
EAMV	Echtes Ackerbohnenmosaik virus
HSV	Hop stunt viroid
LMV	Lettuce mosaic virus
MDMV	Maize dwarf mosaic virus
PDV	Prune dwarf virus
PEMV	Pea enation mosaic virus
PGMV	Peanut green mottle virus
PLRV	Potato leaf roll virus

PMV	Peanut mottle virus
PNRV	Prunus necrotic ringspot virus
PRD	Peach rosette and decline disease
PSTV	Potato spindle tuber viroid
PSV	Peanut stunt virus
PVX	Potato virus X
PVY	Potato virus Y
RCMV	Red clover mosaic virus
RMV	Ryegrass mosaic virus
RNMV	Rice necrosis mosaic virus
RRV	Raspberry ringspot virus
RTV	Rice tungro virus
SBMV	Southern bean mosaic virus
SCMV	Sugar cane mosaic virus
SMV	Soybean mosaic virus
SqMV	Squash mosaic virus
TAV	Tomato aspermy virus
TBSV	Tomato bushy stunt virus
TEV	Tobacco etch virus
TMV	Tobacco mosaic virus
TNV	Tobacco necrosis virus
ToYMV	Tomato yellow mosaic virus
TRSV	Tobacco ringspot virus
TRV	Tobacco rattle virus
TSV	Tobacco streak virus
TSWV	Tomato spotted wilt virus
TVMV	Tobacco vein mottle virus
TYMV	Turnip yellow mosaic virus
WCMV	White clover mosaic virus
WMV	Watermelon mosaic virus
WSMV	Wheat streak mosaic virus
WTV	Wound tumour virus

CHAPTER 1

Why Study the Biochemistry of Virus-Infected Plants?

1.1. INTRODUCTION: VIRUSES AS PATHOGENS AND 'PETS'

1.1.1. The economic importance and academic interest of viruses

Viruses cause severe losses in most crops. Some species are almost universally infected: until techniques for elimination of viruses from cassava and potato were developed comparatively recently, it was impossible to obtain their potential yields. Chance infections of virus-free crops also can cause immense damage. It has been estimated that the annual loss to virus diseases of all crops in the USA is $1.5 to 2 billion (Bialy and Klausner, 1986). In the UK, TMV caused the loss of up to 25% of the glasshouse tomato crop until effective control measures were developed (Broadbent, 1964). Within a decade, about 10 million citrus trees were lost to CTV in Brazil, and over 100 million cocoa trees to CSSV in Ghana (Gibbs and Harrison, 1976). In the Philippines, 30 million coconut palms are thought to have been killed by a sub-viral pathogen, the cadang-cadang viroid (Randles, 1975). Many further examples could be cited.

Clearly, there is a strong economic argument for increasing the ability to control plant virus diseases. In part, this will depend on an understanding of the fundamental biochemistry of the interaction between a virus and its host plant.

However, it would be misleading to suggest that most scientists have approached their study of virology in such a practical spirit. Plant viruses have an intrinsic fascination. They have served as

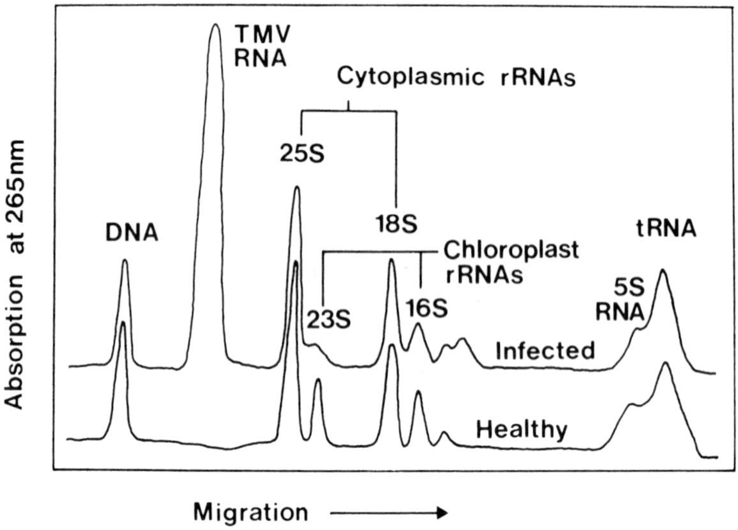

Fig. 1.1. Fractionation by polyacrylamide gel electrophoresis of total nucleic acids extracted from healthy and TMV-infected tobacco leaves, 15 days after inoculation.

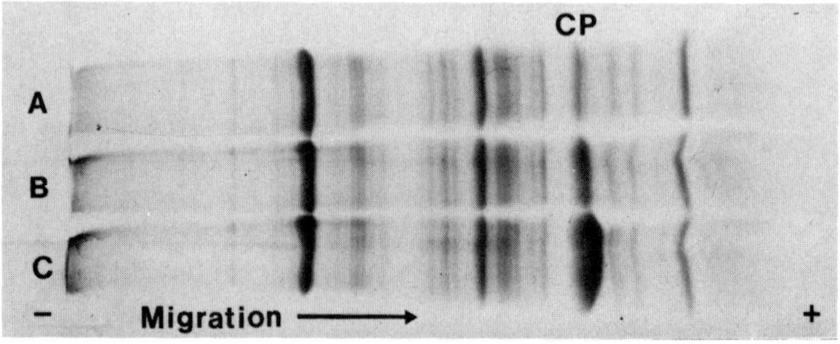

Fig. 1.2. Fractionation under denaturing conditions by polyacrylamide gel electrophoresis of total proteins extracted from TMV-infected tomato leaves, 14 days after inoculation. Tracks A and B show proteins from plants with the Tm-1 gene for TMV resistance in the homozygous and heterozygous forms, respectively. Track C shows proteins from a susceptible plant without Tm-1. CP indicates viral coat protein. Note the accumulation of coat protein in the susceptible plant, and the Tm-1 gene-dosage dependent inhibition of accumulation.

valuable model systems for such diverse topics as the interactions between macromolecules, the genetic code, mutation, and gene expression in higher plant cells. These kinds of investigation are far removed from crop protection, but have yielded much useful information about fundamental problems in biochemistry and molecular biology. Furthermore, knowledge of the processes of virus replication, and the ability to manipulate viral components *in vitro*, is now offering the possibility of using viruses as gene vectors for plant transformations (Hull and Davies, 1983; Siegel, 1985).

The study of the molecular biology of plant viruses has been greatly helped by the ability to purify them in large quantities, and to carry out experiments on topics such as assembly or translation *in vitro*. Determination of the base sequences of viral genomes is giving further insights into virus structure and some aspects of function. In molecular terms, viruses are certainly the most completely understood of the plant pathogens.

In biological terms, however, there are major gaps in our knowledge, and the picture is immensely more complex than when the virus is considered merely as a chemical entity. To illustrate this: it is possible to take apart a virus such as TMV, and to reassemble the protein and RNA components in the test tube to form functional virus. The entire sequence of the TMV genome is now known, but we do not yet fully understand how TMV replicates in the plant. We do not know completely how it causes its myriad effects on the host, including the mosaic from which the name is derived. This book is an attempt to summarize our current knowledge of the plant side of the host-pathogen interaction.

To put the problem in a different, more quantitative light, Fig. 1.1. shows total nucleic acids extracted from a tobacco plant infected with TMV, together with nucleic acids from a healthy plant for comparison. It is clear that the TMV RNA had become the commonest nucleic acid in the infected plant, representing some 75% of the total. Fig. 1.2 shows that likewise, virus coat protein had become the commonest protein in TMV-susceptible tomato. TMV is probably the most successful plant virus in terms of yield; most other plant viruses multiply to much lower levels. However, TMV is useful in showing the maximum demands a virus can make of the host. In a successful infection, TMV nucleoprotein can form up to 10% of the dry weight of the leaf lamina.

A plant virus contains enough genetic information to code for about 4 to 10 proteins. How can it take over the immensely more complex host - containing perhaps 10,000 to 100,000 genes - and turn it into what is virtually a factory for virus production?

1.1.2. The diversity of plant viruses

Plant viruses have been classified into almost 30 groups, and these are increasing as more viruses are described. There is considerable diversity between groups (Francki, 1978; Matthews, 1979). Detailed descriptions of the members of each group are given in Kurstak (1978), Matthews (1981) and Francki _et al_. (1985a; 1985b). Fig. 1.3 illustrates some representatives of various major groups, and Table 1.1 shows the range of diversity in size, composition and complexity over all the groups. It is appropriate to include the viroids, because although they are in some ways quite distinct from viruses, they have some similarities in patterns of replication, and show analogous types of interactions with the host.

The range of total particle mass, from viroid to wound tumour virus, is 1:550.

The genetic material is present in a variety of forms. The range of genomic size, from the smallest viroid to the largest plant virus, is about 1:125. For comparison, the range of genomic size in a very widely-defined plant kingdom is 1:300. This covers from the simplest eukaryote, the fission yeast _Schizosaccharomyces pombe_, to an appropriately representative higher plant, the garden pea _Pisum sativum_.

Satellite RNAs, which depend for their replication on the presence of a helper virus but have no sequence homology with it, also occur in association with some virus groups.

Viroids specify no known proteins, although they might be associated with host-coded proteins _in vivo_. The majority of plant

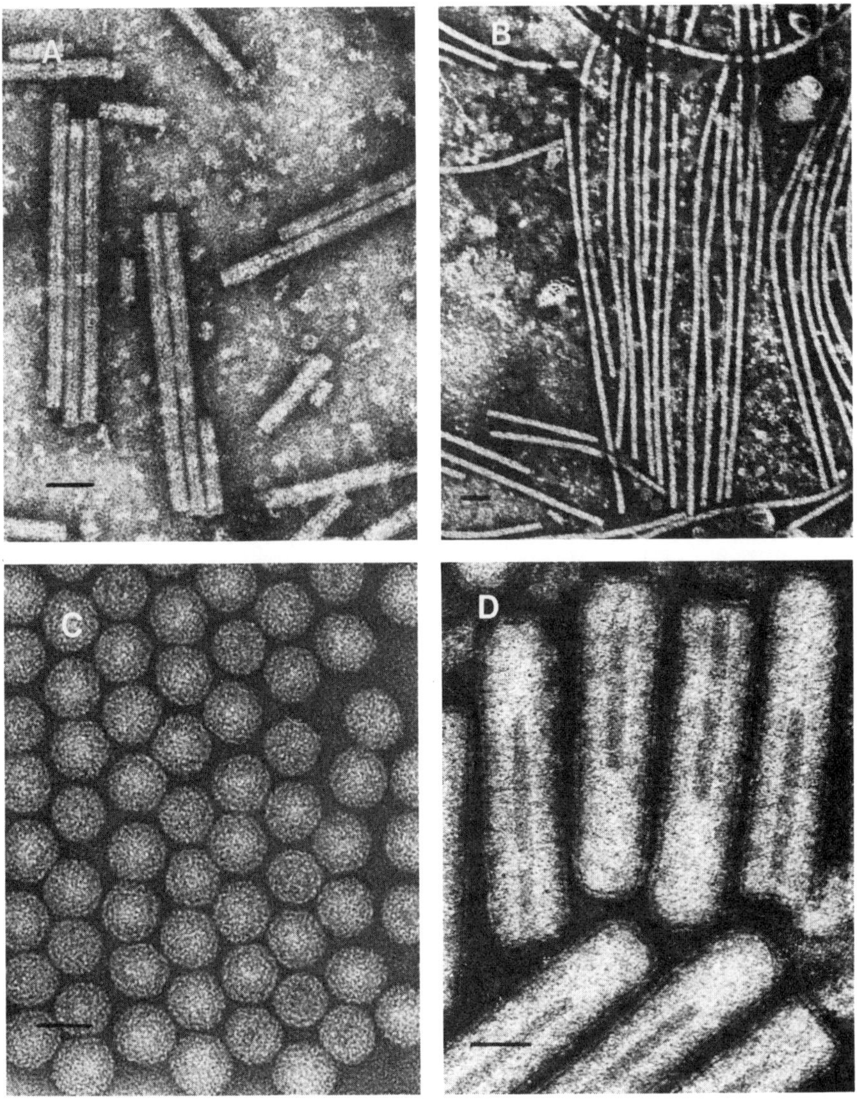

Fig. 1.3. Electron micrographs showing representative plant viruses. (A) Tobacco mosaic virus, with stiff, rod-shaped particles.(B) The long flexuous rod-shaped particles of turnip mosaic virus.(C) The isometric particles of cauliflower mosaic virus.(D) Bullet shaped particles of a plant rhabdovirus, broccoli necrotic yellows virus, with each particle enclosed in a lipoprotein membrane. The scale bars show 50 nm. Micrographs by Colin Clay, National Vegetable Research Station.

Table 1.1. Some features of the compositions of viroids and plant viruses.

Total molecular mass:	125 kDa - 68 MDa
Genomic configurations:	DNA single-stranded DNA double-stranded RNA single-stranded linear RNA single-stranded circular RNA double-stranded (+) or (-) sense where applicable single or multicomponent genome satellite RNAs
Proteins:	absent purely structural structural and catalytic
Lipid:	absent as lipoprotein membranes
Small molecules and ions:	polyamines divalent cations water

viruses have protein, normally of a single type, with a purely structural role as coat protein. The larger particles have several types of structural proteins as well as enzymic activities such as polymerases.

Lipids are absent from most virus groups, but are present in plant rhabdoviruses and TSWV in the form of lipo-protein membranes.

Finally, a few viruses contain polyamines, and most or all rely on divalent cations for structural stability.

It should be clear that viroids and plant viruses are very heterogeneous, and that they might be expected to show various patterns of interaction with their host plants, with different demands and effects on host biochemistry.

Table 1.2. 'How to be a successful virus'. An anthropomorphic approach defining some possible requirements for an existence as a selfish gene.

1. Originate: create a unit of information bearing the message 'Please replicate me in time and space'.

2. Protect the information from damage or corruption.

3. Multiply the information (the viral genome).

4. Multiply the other components, such as coat protein.

5. Assemble components to form progeny units of protected information (virus particles).

6. While performing steps 3 - 5, control host metabolism as necessary, to create conditions favourable for the multiplication and promulgation of the information.

7. Transmit the information through space and time, to further host plants.

8. Evolve the content of the information as necessary, so that the command remains intelligible to the host.

1.2. INTERACTIONS OF VIRUSES WITH HOST PLANTS

1.2.1. How to be a successful virus

A virus is the epitome of the selfish gene, as defined by Dawkins (1976). It is basically a piece of information: the message 'please replicate me'. Some of the more complex viruses also contain information regarding their dissemination in space and time. The success of a virus depends on the frequency with which an organism - in this case a plant - can be directed to obey the message.

It is perhaps useful to adopt an anthropomorphic approach at this stage, as an aid to asking what sort of help a 'new' virus might need, to allow it to pursue this 'lifestyle'. This may give some insight into the range of 'demands' that a virus might have to make of the host in order to be successful. Table 1.2 is a list of possible areas where a host and a virus might interact, represented as good advice to a young virus embarking on a career in (something else's) life.

There are many theories of the origin of viruses, and little evidence for any of them. Given the range of diversity (Table 1.1), it is reasonable to suggest that plant viruses and viroids may have originated in several distinct ways; i.e. they are polyphyletic. The origin might be exogenous to plants, in which case one is left with the problem of how a piece of information from elsewhere could ask the plant: 'please replicate me'. Or the origin might be endogenous, in which case we are considering an established piece of information, already being replicated by some plant system, which escapes from normal controls. In the endogenous case, therefore, the plant contributes the fundamental matter of existence to the virus. This might occur at several levels.

There are similarities in base sequence between viroids, the U1 small nuclear RNAs (sn RNAs) involved in messenger RNA processing, and processing sequences on the mRNA precursors. It has therefore been suggested that viroids might represent escaped introns, the internal sequences in the mRNA precursor which are excised during processing (Diener, 1981). An alternative possibility is that viroids might represent escaped snRNAs.

In a related but more evolved model, Zimmern (1982) has suggested that viroids and RNA viruses might have been derived from a system involved in the interchange of genetic information between eukaryotic cells. One form of the model envisages that a molecule normally involved in a 'signalling' function, and thus with an implied specificity of transport, is accidentally coupled with polymerase-recognition and transcription-initiation sequences, which allow it to short-circuit normal controls over its synthesis. The presence of RNA-dependent RNA polymerase activity in healthy plants - an enzyme with an as yet undiscovered function, as explained in Chapter 2 - might be one route to the origin of viruses.

Similar arguments may be applied at the DNA level. For example, a transposable element which escaped from normal controls might be involved in the origin of DNA viruses. By analogy with retroviruses of mammals, which have manifestations as both RNA and DNA, it is also possible that plant RNA viruses might ultimately have had a DNA source. The general lack of sequence complementarity between RNA viruses and the DNA of the host is not an obstruction to this argument, given the high mutational rate of RNA genomes (Reanney, 1982).

This leads to stage 2 of Table 1.2. Viruses protect their information (genomes) from nucleases by encapsidation in coat proteins, and in some cases by being double-stranded. Encapsidation also confers protection against physical forces such as irradiation. There is no evidence that RNA viruses can use any kind of host-specified editing and correction mechanisms to maintain the fidelity of their genomic information. Protection against corruption of the information from errors in copying seems to be largely at the level of producing high numbers of progeny, and sometimes utilizing the statistical protection conferred by having a split genome (Reannay, 1982). The requirements to produce large numbers of copies of the genome, and large amounts of protecting coat protein, however, make clear demands on the host.

Stages 3 to 6 of Table 1.2 are where the clearest and most direct interactions between plant and virus occur. Detailed consideration of the biochemistry forms the substance of subsequent Chapters. However, it is appropriate here to give brief consideration to some more general points arising from stages 7 and 8.

The strategy adopted for transmission of virus to the next host (stage 7) creates some interesting constraints on stages 3 to 5 of Table 1.2. There are two extremes. A virus which is mechanically transmitted, such as TMV, is dependent on random events. This strategy demands a high level of stability, to allow persistence in the environment, and a high level of virus multiplication, to increase the chance of the progeny contacting a new host. This makes further, quantitative demands on the previous host.

In contrast, a virus which is transmitted by a specific vector, such as an insect feeding on the host species, has a much more secure and specific pathway for spread. In this case there is less requirement for the production of large amounts of virus particles, with a consequent reduction in demand on the host. This may allow the evolution of a more subtly-regulated interaction between the host and the virus. For example, viruses which are transmitted by insects feeding only from the phloem, can be restricted to the phloem in infected plants (Kubo and Takanami, 1979).

Viruses transmitted by host-specific vectors may be considered as parasitizing an 'extended host', which may also include the vector organism. The fact that some rhabdo- and reoviruses of plant and animal origin are structurally very similar, and that some can multiply both in plants and the insect vectors, also has a bearing on possible origins. Some plant viruses could have developed from animal DNA or RNA sequences which escaped control.

The final point in Table 1.2 is evolution, and here it is important to emphasize that agriculture and plant breeding have most probably altered the environment for viruses, and the selection pressures on their evolution. Most virus studies involve crop species, which have been bred and selected for agronomically useful attributes. The breeding may have distorted host-virus interactions. Some of the biochemistry in this book might therefore be man-made, and different from the naturally-evolved relationships between viruses and wild species (Tomlinson et al., 1970). The most striking selection pressure on virus evolution has resulted from the deployment of resistance genes in breeding. The evolution by the virus of corresponding virulence, the ability to overcome particular resistance genes, is considered in Chapter 7.

1.2.2. How virus-plant interactions are approached in this book

The following Chapter will discuss the most direct area of involvement of the host with the virus, by considering the role of host-specified components in virus multiplication. The succeeding Chapters then deal with major areas of host metabolism, i.e. nucleic acids, proteins, respiration and photosynthesis, and consider how virus infection affects them directly and indirectly. These are followed by a discussion of the biochemistry of a particular class of host response, namely resistance. This is the most important means of controlling virus diseases in agriculture, and the mechanisms involved have been studied intensively. Because of this, and the intrinsic interest of resistance mechanisms, the biochemistry is dealt with in some detail. The next Chapter deals with how viruses influence host growth and development, and how they cause visible symptoms. This Chapter draws information from, and relates to, all the preceding Chapters. The final Chapter summarizes progress in understanding the biochemistry of plant-virus relationships, and makes some predictions of future developments.

CHAPTER 2

The Role of the Plant in Virus Replication

2.1. INTRODUCTION

Viruses, by nature and by definition, are unable to multiply without the assistance of a living cell. It is central to the definition that the cell provides nucleoside triphosphates and amino acids as precursors of viral nucleoprotein, any other materials involved in viral structure such as membranes, polyamines or ions, an energy-generating system, and a means of translating viral messenger RNAs. No virus contains these abilities as part of the particle, or specifies them completely within the host cell. Other areas of the viral replicative cycle which may rely on a host-specified factor are less well understood. They include uncoating; replication of the viral genome; assembly and transmission. The links between virus multiplication and pathways of energy supply are considered in relation to respiration and photosynthesis in Chapter 5.

Fig. 2.1 shows how some of the potential areas of host involvement in multiplication may be related, for a stylized representative virus. 'Multiplication' is defined very broadly in this context to include all processes involved in the increase in virus numbers in space and time. The aim of this Chapter is to review the evidence for the involvement of host functions. Aspects of replication which are purely viral, such as the different strategies of genome expression, virus-coded transcriptases and the biochemistry of self-assembly, have been dealt with more extensively by others (Kurstak, 1981;

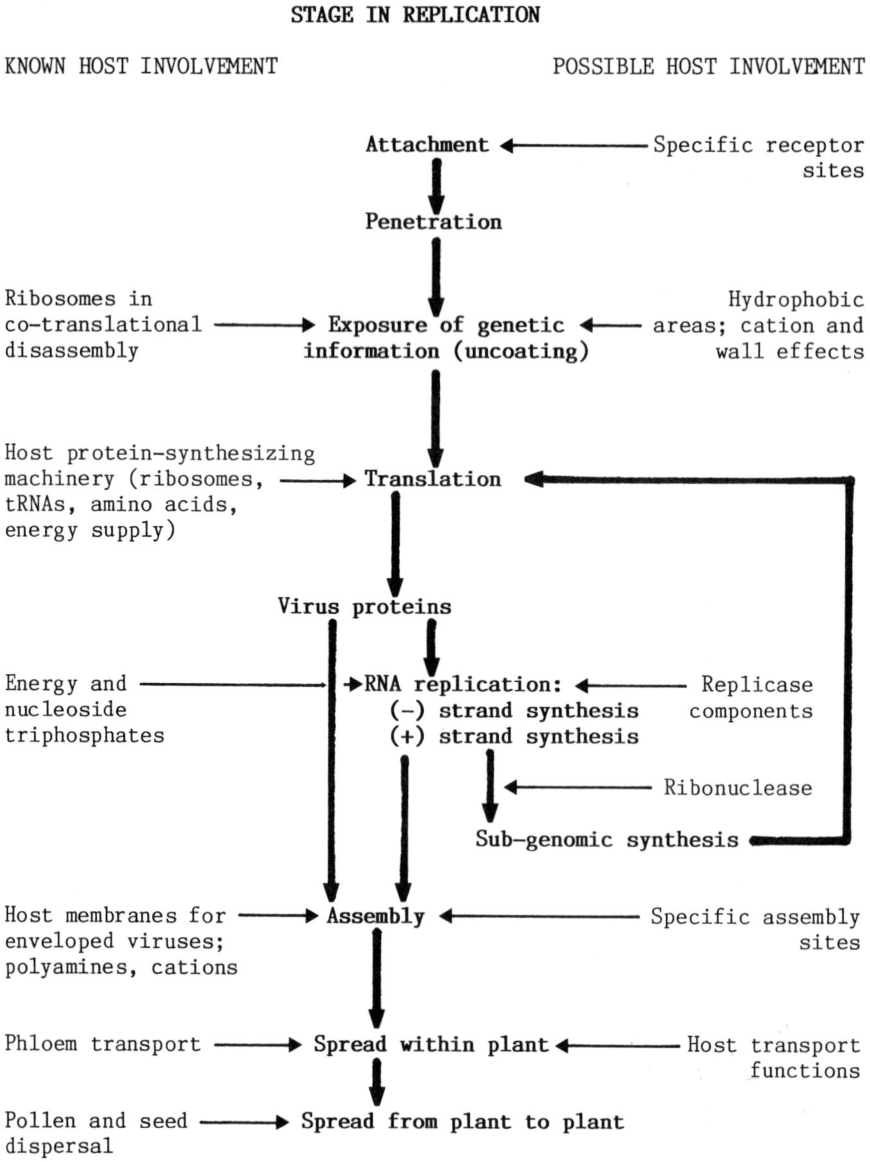

Fig. 2.1. Known and possible involvement of host-specified functions in the multiplication of a representative virus with a positive-sense RNA genome. Virus-controlled functions are shown by heavy arrows, and host functions by light arrows.

Davies and Hull, 1982; Butler, 1984; Davies, 1985a; 1985b; Francki, 1985; Francki et al., 1985a; 1985b). These purely viral aspects will be considered here only where there is interaction with host factors.

2.2. THE INITIATION OF INFECTION

To initiate an infection, a virus must make contact with the host; penetrate to a suitable site to establish infection, and expose its genetic information for expression by the host translational machinery. The last step may involve partial or complete uncoating of the nucleic acid.

2.2.1. Does the host provide specific receptor sites?

The overwhelming weight of the evidence suggests that it does not (Novikov and Atabekov, 1970; Atabekov, 1975; Shaw, 1985). The problem has received most attention in the context of the possible role of specific receptor sites in determining host range, and is considered more fully in this light in Chapter 6.2.2. It appears likely that for mechanically-transmitted viruses, the wound produced by the event causing inoculation is enough to allow virus entry. Thomas and Fulton (1968) suggested that the non-specific resistance of the tobacco line T.I. 245 to infection by a number of viruses was due to its low number of ectodesmata. For vector-transmitted viruses where entry to the susceptible tissue can be direct, there is also no need to postulate any specific attachment site.

It is interesting here to contrast viral infections of plant cells on the one hand, and animal and bacterial cells on the other. For the latter two, there is very good evidence that cell surfaces do contain virus-specific recognition sites (e.g. Bukrinskaya, 1982; Wunner, 1982; Reanney and Ackermann, 1982). The difference may reflect the ways in which the surfaces of animal and bacterial cells are exposed to viruses in nature, whereas plant cell membranes are normally inaccessible because of the cell wall. This imposes a requirement for wounding, either mechanically or by a phytophagous vector. Plant systems can be made to resemble animal systems experimentally by preparation of protoplasts, but studies of infection processes do not suggest any requirement for virus-attachment sites on the surface of the plasma membrane (e.g. Fukunaga et al., 1981; Motoyoshi and Oshima, 1979).

2.2.2. The role of the host in uncoating

Uncoating, or disassembly, of plant viruses has been extensively studied in vitro, as one means of probing the nature of virus structure and protein-nucleic acid interactions. This form of uncoating generally has required rather extreme conditions such as high pH (e.g. Perham and Wilson, 1978) or ionic detergents (e.g. McCarthy et al., 1980). Although the processes observed might bear some resemblance to uncoating in vivo, the latter must involve much milder conditions, and therefore presumably must also involve other agents.

Studies of the kinetics of uncoating of the RNAs of rod-shaped and isometric viruses in the plant have suggested that the process is rapid, with detectable or even complete release occurring within minutes (Shaw, 1970; Matthews and Witz, 1985). Shaw's earlier (1969) demonstration that release of TMV RNA also occurred in non-hosts would tend to argue against host-specific factors being involved in uncoating. However, there is no reason why host specificity should be expressed at this stage, or why non-hosts should not contain some general factor promoting uncoating of viruses. A further comment on these experiments is that the majority of the uncoated particles failed to initiate infection; there might in fact be two or more pathways of uncoating, with most particles entering a non-specific, degradative pathway.

There are various theories on how viruses might be uncoated after making contact with the host. Interactions have been shown between the cell wall and TMV (Kassanis and Kenten, 1978) and TRV (Gaard and De Zoeten, 1979), but these probably lead to non-specific degradation without initiation of infection. Durham and Hendry (1977) drew attention to the possible importance of Ca^{2+} ions in TMV structure, and suggested that loss of these cations from particles was a necessary but not a sufficient condition for disassembly. Whether such loss would occur during penetration of the cell is unestablished. Others have noted the importance of hydrophobic interactions in the binding of TMV coat protein subunits, and have suggested that the hydrophobic environment of membranes would promote disassembly by interfering with this stabilizing force (Kiho et al., 1979). Interactions of TMV coat protein with lipid membrane systems have been demonstrated in vitro (Banerjee et al., 1981), but the significance for uncoating in vivo has not been established.

The most interesting evidence for a role of host components in uncoating has come from recent work by Wilson and collaborators. When TMV particles were added to cell-free protein-synthesizing systems based on rabbit reticulocytes (Wilson, 1984a) or wheat germ (Wilson, 1984b), they served as highly efficient templates for protein synthesis (Fig. 2.2). The polypeptides which were produced resembled those produced by translation of free TMV RNA, but were not completely similar. Examination of protein synthesizing systems containing TMV particles showed particles of various lengths, with ribosomes attached to the protein-free RNA tails (Fig. 2.3). These complexes have been termed 'striposomes' and the process in which they are formed 'co-translational disassembly'.

Before TMV particles could serve as efficient templates in cell-free protein synthesizing systems, it was necessary to wash briefly under very mildly alkaline conditions. Wilson and Shaw (1985) suggested that this was sufficient to remove a small amount of protein from the 5' end of the TMV particle, to expose an early AUU site for ribosome binding. Subsequently, translation of the RNA - an energy requiring process - leads to sequential displacement of coat protein subunits as the first ribosome moves along the template.

These results of course refer to in vitro systems, but very recently Shaw et al. (1986) have demonstrated cotranslational disassembly in vivo. Striposomes from plants behaved similarly on cesium sulphate gradients to those prepared in vitro, and some of the complexes appeared similar in the electron microscope. Some further, indirect evidence for the occurrence of striposomes in vivo comes

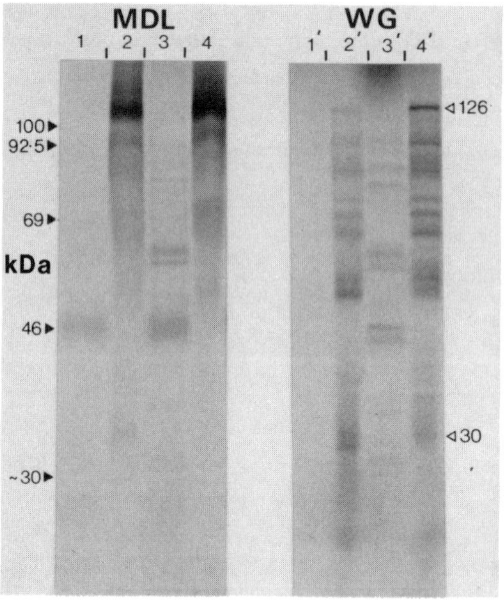

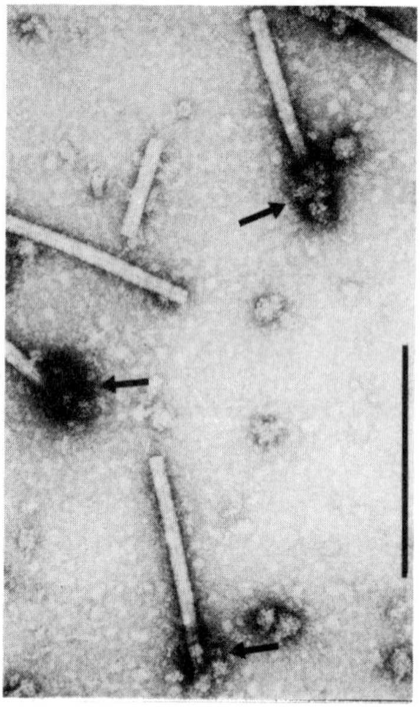

Fig. 2.2. (Left) Autoradiographs of polyacrylamide gel fractionations of radiolabelled proteins synthesized in vitro in rabbit reticulocyte lysate (MDL) and wheat germ extract (WG). Lanes 1 and 1'; control incubations with no added mRNA. Lanes 2 and 2'; TMV RNA. Lanes 3 and 3'; TMV rods stored at pH 7.0. Lanes 4 and 4'; TMV rods treated briefly at pH 8.0 to remove a small number of coat protein subunits at the 5'-end. From Wilson (1984b), by permission of the author and Academic Press.

Fig. 2.3. (Right) Electron micrograph of 'striposome' complexes formed during incubation of TMV particles with a rabbit reticulocyte lysate. Arrows show clusters of ribosomes disassembling TMV particles. The bar shows the full-length (300nm) of TMV particles. From Wilson (1984a), by permission of the author and Academic Press.

from earlier work by Kiho et al. (1972). They found that
polyribosomes formed during the early stages of TMV infection
appeared to be associated with partly-uncoated virus particles.
Hayashi (1977) also reported partial uncoating. This leaves, however,
the problem of how the coat protein subunits at the 5' end might be
removed in vivo to expose the ribosome binding site. Wilson and Shaw
(1985) speculated that this could involve the Ca^{2+} or hydrophobic
interaction models. Both should be open to experimental test.

Does this model for active participation of the host in uncoating
apply to other viruses? Brisco et al. (1985) showed that particles of
SBMV, an isometric virus, can act as templates for protein synthesis
in the rabbit reticulocyte system. Again, dialysis against a mildly
alkaline buffer was required for activity. This appeared to involve
removal of a few coat protein subunits to open a 'molecular trapdoor'
through which the RNA could be removed during translation. The capsid
did not collapse, and the RNA was not exposed until translation had
commenced. Wilson (1985) and Wilson and Shaw (1985) mention a few
other viruses which seem to support the model, and two potexviruses
which do not. The latter observation is intriguing: it might suggest
that there is another, quite different mechanism of uncoating, which
does not depend on co-translational disassembly. But a trivial
explanation is also possible, in that the correct conditions for in
vitro expression may simply not have been found.

One of the attractions of the co-translational disassembly model is
that the RNA is protected until the last moment from degradation by
ribonucleases. Furthermore, it appears that the genes for TMV
proteins which may be involved in the replicase are located at the 5'
end; a similar argument is made for other viruses (Kamer and Argos,
1984; Franssen et al., 1984b). Early expression of the replicase
polypeptides would facilitate the next phase of multiplication.

Wilson and Watkins (1985) found that co-translational disassembly
of TMV particles appeared to be inhibited when the ribosomes reached
the assembly initiation site, still some 1030 bases distant from the
3' end (Goelet et al., 1982). Coat protein-RNA binding forces are
likely to be particularly strong in this region. Wilson and Watkins

(1985) speculated that the possible replicase, which had by then been translated from the 5' end, might bind to the 3' end and complete the uncoating in a 3' to 5' direction. This process, which they termed 'co-transcriptional disassembly', could be purely a matter for virus-specified components, unless the replicase activity involves host-coded subunits.

2.3. TRANSLATION

2.3.1. What types of ribosomes synthesize viral proteins?

Apart from their proposed role in uncoating, the plant ribosomes synthesize viral proteins which are probably involved in replication, assembly of the virion and transport of the progeny virus particles. Plant cells contain cytoplasmic (80S), chloroplast (70S) and mitochondrial (78S) ribosomes. The last have not been considered in the context of viral protein synthesis.

The possibility that chloroplast ribosomes might be involved is of interest, in view of the production of visible disease symptoms which must involve chloroplast changes, and the known association of viruses such as TYMV with chloroplasts (Matthews, 1973). However, experiments with specific inhibitors of protein synthesis on cytoplasmic (cycloheximide) and chloroplast ribosomes (chloramphenicol) have suggested that only cytoplasmic ribosomes are involved in synthesis of the proteins of TMV (Paterson and Knight, 1975), CPMV (Owens and Bruening, 1975) and PVX (Kraev and Diden, 1982). Interpretation of experiments involving such differential inhibitors, especially chloramphenicol, can pose problems, as emphasized by Ellis (1981a). A useful internal control of specificity has been to follow effects on the large and small subunits of ribulose 1,5-bisphosphate carboxylase, which are synthesized in the chloroplast and cytoplasm respectively. This approach was adopted by Owens and Bruening (1975). Overall, the lack of evidence for any involvement of chloroplast ribosomes in synthesis of viral proteins is consistent with the rapid degradation of chloroplast ribosomal RNA observed in some virus infections (see Chapter 3.2).

Virus protein synthesis also depends on host transfer RNA and amino acid supply; little is known about possible changes in transfer RNA after infection, but some effects on amino acid concentrations and activation are described in Chapter 4.6.

2.3.2. Do host factors control the specificity or relative amounts of virus proteins synthesized?

In vitro translation of viral RNAs can lead to production of a larger total amount of polypeptide sequence than could be coded for by the nucleic acid on a straight ratio of three base residues per amino acid (e.g. Wilson and Glover, 1983). Furthermore, the particular proteins synthesized in vitro may differ from those synthesized in vivo (Goldbach et al., 1981; Wilson and Glover, 1983), although for other viruses the in vitro translation appears to be 'correct' (reviewed in Davies and Hull, 1982). This raises the question of whether, in the former case, host components have any direct role in control of translation of the viral genome. At present, this possibility cannot be excluded, but the weight of the evidence suggests that the complexity of translation of viral sequences, and the mis-match between in vitro and in vivo results, are either a product of virus-defined mechanisms, or can be explained by non-specific effects.

Synthesis of some viral proteins involves translation of sub-genomic messengers (e.g. Higgins et al., 1976; Gonda and Symons, 1979; Sulzinski et al., 1985). A possible role of the host in production of sub-genomic messengers will be assessed in the next section. Other viral RNAs are translated to produce a polyprotein which is then cleaved to the functional products (e.g. Goldbach and Krijt, 1982; Franssen et al., 1984a). The specificity of this process may be determined entirely by the virus: Goldbach and Krijt (1982) showed that the large RNA of CPMV codes for a protease which cleaves the polyprotein produced by translation of the smaller genomic RNA. The protease was shown to be highly specific in that it did not cleave polyproteins of other comoviruses. Wilson and Glover (1983) have suggested various mechanisms which might be involved in synthesis of virus—coded polypeptides in vitro, which are not detectable in vivo.

2.4. TRANSCRIPTION

2.4.1. Replicases and transcriptases: some job descriptions

The number of detailed types of transcriptional activity required for full expression and replication of the genetic information of viruses and viroids is probably quite large. A priori, several functions can be envisaged. Also, several types of constraint can be postulated to modify functions or place particular requirements on transcriptional activities. Finally, the different forms of pathogen – positive and negative sense RNA viruses, double stranded RNA viruses, single and double stranded DNA viruses, and viroids – are likely to require different forms of transcriptional activity during their replicative cycles.

2.4.2. Host components in the replication of positive sense RNA viruses

The enzyme(s) involved in replication of RNA viruses with (+) sense RNA genomes are required to copy the (+) sense genomic RNA to produce a complementary (−) sense strand, which is then copied to produce (+) strand progeny. To fit observations, there is a requirement to produce a much larger amount of (+) sense progeny than of (−) sense complementary strand (Nassuth et al., 1983b), although this balance could be regulated at levels other than that of the replicase (Nassuth and Bol, 1983). Finally, there is a requirement to produce the sub—genomic messengers by transcription of (−) sense RNAs; there are several possible routes to this.

Research on plant virus replicases has been guided to some extent by knowledge of bacteriophage and animal virus systems. Both virus- and host—specified components are involved in the RNA replicases of, for example, phage Qβ (Jockusch, 1974; Kamen, 1975) and poliovirus (Dasgupta et al., 1980). For plant viruses, most of the evidence for a virus—coded replicase or replicase components is indirect, or is argued by analogy. Thus the RNA sequences of plant viruses in several groups have regions which can be aligned with the sequence coding for the known replicase of poliovirus. The degree of sequence homology is sufficient to suggest that the aligned sequences of the plant viruses must specify an enzyme with similar activity (Kamer and Argos, 1984; Franssen et al., 1984b). Also, evidence from experiments with multicomponent viruses has shown that single components (Goldbach et al., 1980), or a pair of components which does not represent the complete genome (Nassuth and Bol, 1983), can be replicated and expressed in plants without the presence of the remainder of the genome, but that other combinations representing incomplete virus genomes cannot. This suggests strongly that the virus does contribute to the replicase, and excludes the possibility that replication of a viral RNA is simply a result of its association with an enzyme system specified by the host alone.

Experimental work on possible enzyme activities involved in replication of plant viruses with (+) sense RNA genomes began to be published in the early 1970s, and the position quickly became very confused. This was partly because properties which can be proposed for a true replicase were, in most cases, not fulfilled. A true replicase and transcriptase complex would be expected to copy (+) to (-), and (-) to (+); to be dependent on endogenous or added template; to show some degree of specificity for the homologous viral RNA; to be capable of initiation, elongation, termination and release; and to be able to synthesize full-length and sub-genomic viral RNA molecules. In very few studies have all, or indeed any, of these properties been demonstrated.

The situation also became confused when it was discovered that healthy plants contain RNA-dependent RNA polymerase activities which are able to synthesize RNA in vitro (reviewed in Duda, 1976). Again, these enzymes fail to show most of the features required of a full replicase activity; their relationship to the 'true' replicase is still much in doubt. Further confusion was introduced by the discovery that RNA-dependent RNA polymerase activities could exist in 'soluble' and 'membrane-bound' forms, sometimes with different properties. Finally, enzyme preparations have frequently been found to contain a template-independent poly(U)-polymerase activity, which co-purifies with RNA polymerase in the early stages of most extraction procedures (Brishammar, 1976; Chifflot et al., 1980). These are an especial problem in that most workers have used incorporation of radioactive UTP to measure polymerase activity.

The enzymology of RNA synthesis has been intensively studied for eight main positive-strand RNA viruses: CMV, BMV, CCMV, CPMV, AMV, TYMV, TNV and TMV. The level of understanding varies considerably from system to system. This section will consider representative examples, asking which host-specified activities, if any, are involved in replicase and transcriptase functions, and how host-specified components might interact with virus-specified replicative functions.

Cucumber mosaic virus. Both soluble and membrane-bound RNA-dependent RNA polymerase activities are induced by CMV infection of cucumber seedlings; the solubilized 'bound' enzyme appears to be similar in properties to the soluble enzyme (Kumarasamy and Symons, 1979; Gill et al., 1981). The most highly purified enzymes so far have contained up to nine polypeptides, but the majority of these were also present in preparations from uninfected plants. The infection-specific bands had molecular masses of 100, 110 and 35 kDa, although the first was by far the commonest, and co-purified with the polymerase activity. Surprisingly, these proteins did not correspond to the translation products of the CMV genome (Gordon et al., 1982), when compared by two peptide mapping procedures. This suggests that the polymerase activity as extracted was entirely host encoded.

These enzyme preparations have been referred to in the literature as CMV 'replicases', but have shown no specificity for CMV RNA (May et al., 1969), and the products of transcription have generally been small and heterogeneous (Kumarasamy and Symons, 1979). Recently, however, Jaspars et al. (1985) demonstrated synthesis of full-length copies of the four CMV RNAs. The product was overwhelmingly positive sense, appeared to involve elongation of strands which had already been initiated *in vivo*, and was not released from the replicative complex as free single-stranded RNA. The small heterogeneous products were also made, and these were shown to be (+) and (-) sense copies of plant and viral RNAs. Jaspars *et al.* (1985) discussed the extent to which their *in vitro* assay system might be representative of the full CMV replicase activity, and speculated that the host-coded 100 kDa polymerase might be involved in initiation of synthesis of CMV RNAs *in vivo*. The possible role of the virus-coded non-coat proteins in CMV RNA synthesis remains in doubt.

Cowpea mosaic virus. This bears some similarity to the CMV results, in that the 'replicase' *in vitro* synthesizes full length virus RNAs and also small heterogeneous RNAs (Dorssers *et al.*, 1983a). These authors and Van der Meer *et al.* (1984) suggested that the host-coded RNA polymerase was responsible for synthesis of the small RNAs but not involved in virus RNA synthesis; Dorssers *et al.* (1984) presented evidence suggesting that a virus-coded 110 kDa polypeptide was the polymerase responsible for elongation of virus RNAs. However, the former conclusion was challenged by Jaspars *et al.* (1985) on the grounds that the *in vitro* assay systems did not represent a complete viral replicase function. These assay systems could not therefore provide evidence to exclude a role for the host-specified RNA polymerase in virus replication.

Turnip yellow mosaic virus. The replicase complex in Chinese cabbage leaves is associated with the outer membrane of the chloroplast (Lafleche and Bove, 1971), and contains viral RNA template when prepared from infected leaves. The product was shown to be (+) sense viral RNA. However, a template-free enzyme could be prepared, and this used viral (+) sense RNA to produce a (-) sense product of up to full length (Mouches *et al.*, 1974). The association of the replicase with the chloroplasts, and the fact that in Chinese cabbage the host-encoded RNA-dependent RNA polymerase activity is entirely soluble (Astier-Manifacier and Cornuet, 1971), made it comparatively easy to separate the replicase from other polymerase activities in

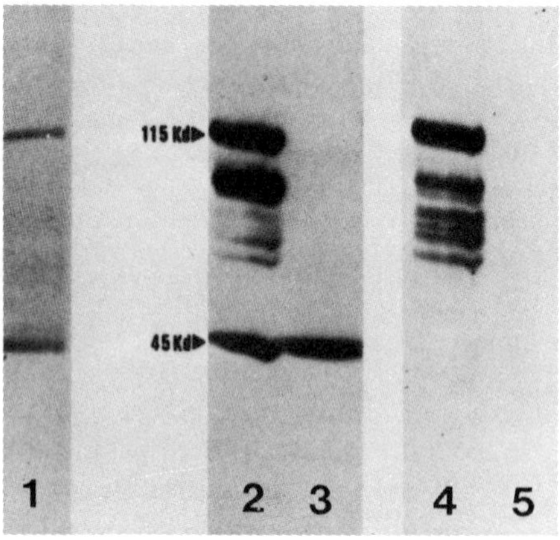

Fig. 2.4. Electrophoresis of heat-denatured, purified TYMV replicase on polyacrylamide gels containing sodium dodecyl sulphate, showing identification of the host and virus-coded subunits. Track 1: purified replicase stained with Coomassie blue, showing two main bands. Tracks 2 and 3: replicase preparations from infected and healthy plants respectively. After electrophoresis, the proteins were blotted to nitrocellulose filters and incubated with anti-replicase serum. The immunocomplexes formed were detected by labelling with radiolabelled protein-A. Tracks 4 and 5, proteins from infected and healthy plants respectively, blotted and incubated with anti-replicase serum as above, then incubated with radiolabelled _in vitro_ translation products of TYMV RNA. From Mouches _et al._, 1984, by permission of the authors and Academic Press.

this system. In most other plants, the host RNA-dependent RNA polymerase activity occurs in both soluble and variously-defined membrane bound or particulate forms (e.g. Evans _et al._, 1984; White and Dawson, 1978).

The TYMV replicase has been purified to near homogeneity, and on sodium dodecyl sulphate gel electrophoresis showed two main bands of molecular weights 115 kDa and 45 kDa (Fig. 2.4, Track 1). Mouches _et al._ (1984) prepared an antibody against the purified replicase, and showed that this inhibited incorporation by about 50% _in vitro_. Gels of replicase preparations from healthy and infected plants were 'blotted' to nitrocellulose paper ('Western' blotting), and incubated with the anti-replicase serum. Immunoglobulins specifically bound to replicase proteins on the blot were then detected using radiolabelled protein A. The replicase preparation from infected leaves showed radioactivity associated with the 115 kDa and 45 kDa bands, as well as with some minor bands. Only the 45 kDa band became radioactive in the healthy leaf extract (Fig. 2.4, Tracks 2 and 3). This suggests that the 45 kDa protein is host encoded. It also suggests that the 45 kDa protein is constitutive of healthy plants, rather than being induced by infection.

When blots were incubated with anti-replicase serum as before, then with radiolabelled in vitro translation products of TYMV RNA, only the 115 kDa and minor bands became radiolabelled in proteins from infected plants, and no bands became labelled in proteins from healthy plants (Fig 2.4, Tracks 4 and 5). This suggests that the 115 kDa band was virus-coded.

The TYMV replicase is one of the best examples where host- and virus-encoded functions have been demonstrated. However, it should be noted that the replicase activity as purified in vitro did not carry out the full range of functions specified above; the true replicase in vivo might be more complex and contain other subunits of host or viral origin.

Brome mosaic virus. The replicase has been partially purified. Non-ionic detergent treatment was critical for solubilization (Bujarski et al., 1982). It was suggested that the determinant of template specificity was intimately associated with the membrane, and was lost if more rigorous membrane-disrupting agents were used. Endogenous template was removed by treatment with micrococcal nuclease (Miller and Hall, 1983). The partially purified enzyme showed a strong preference for BMV RNA as template. It contained a 110 kDa polypeptide which was shown to be coded by the virus RNA 1. The possible role of host-encoded polypeptides in the complete replicase does not seem to have been established.

It will be clear from these examples that many questions remain to be answered about replicases. In some cases, there is evidence for an involvement of host-specified components. The evidence for virus-coded functions is variable. The host RNA-dependent RNA polymerase has been intensively studied, but to date there is little evidence to implicate it in viral RNA replication, and much that suggests that the polymerase and the replicase are separate activities (Dorssers et al., 1983b). The host RNA-dependent RNA polymerase has, to some extent, obscured the search for the true replicase. This is compounded by the difficulties in separating the two activities in most assay systems (e.g. White and Dawson, 1978), and by the large increase in RNA-dependent RNA polymerase activity which accompanies infection in many host plants (Ikegami and Fraenkel Conrat, 1978a; 1978b).

This leaves the question of what role the host RNA-dependent RNA polymerase may perform. The enzyme appears to be present in detectable levels in healthy plants of a number of species, with the exception of cucumber seedlings (Gill et al., 1981). If the enzyme does something important, how can cucumber seedlings do without it?

The products of the enzyme would be expected to be double-stranded (ds) RNAs. These can be found in plants, but only in exceedingly small amounts (Ikegami and Fraenkel Conrat, 1979). This tends to argue against any function of the enzyme in RNA copying, for example in messenger RNA amplification.

The increase in enzyme activity after infection might suggest a role in virus replication, probably in a modified form or with additional host or virus-specified components. Alternatively, the increase might be a response to stress in general terms, and it would be interesting to measure enzyme activities after different kinds of stress. This apparently has not been done.

Finally, it may be that the ability of the enzyme to incorporate NTPs into RNA in vitro might not be a true reflection of its activity in vivo: it might have other functions. Further examination of the enzyme-associated template would also be of interest.

Synthesis of sub-genomic RNAs might involve the same replicase complex as in synthesis of full-length viral RNAs, or a modified form. Miller et al. (1985) suggested that BMV sub-genomic RNAs were synthesized by internal initiation on (-) sense genomic RNA. In contrast, Dawson and Dodds (1982) found a series of sub-genomic ds RNAs in TMV- and CCMV-infected plants, although the kinetics of incorporation of radioactive precursor into these ds RNAs were not fully consistent with an involvement in synthesis of subgenomic mRNAs. In CMV-infected cowpea protoplasts, the kinetics of synthesis of RNA-3 (genomic) and RNA-4 (sub-genomic) suggested that RNA-4 is derived from RNA-3 by nucleolytic cleavage, and not by transcription of a (-) sense RNA-4 (Gonda and Symons, 1979). This cleavage presumably depends on a host ribonuclease.

Addition of polyadenylic acid (poly(A)) sequences to the RNAs of certain viruses is probably dependent on the host poly(A) polymerase. Hari (1980) found an increase in enzyme activity before symptom appearance in tobacco leaves systemically infected with TEV.

2.4.3. Replication of viruses with other types of RNA genomes

The plant rhabdoviruses, with large, single-stranded (-) sense genomes, and the reoviruses, with multicomponent, double-stranded RNA genomes, contain transcriptases in the virus particles. The WTV (reovirus) transcriptase has been shown to synthesize single-stranded RNA copies of all 12 genomic RNAs (Reddy et al., 1977; Nuss and Peterson, 1981). Presumably these transcripts act as monocistronic messenger RNAs in vivo. The virions have also been shown to contain the guanyl-transferase and methylating enzymes necessary to modify the 5'-ends of the transcripts (Rhodes et al., 1977). Thus expression of the WTV genome, in terms of synthesis of suitably capped mRNAs, appears not to depend on host components.

The virion transcriptase which has been found in certain, though not yet all, plant rhabdoviruses, requires removal of the particle membrane or partial dissociation of the virion for expression of activity in vitro (Francki and Randles, 1972; Randles and Francki, 1972; Toriyama and Peters, 1980). The transcriptase products have been reported to be complementary to the genomic RNA but to be of much lower molecular weight, with a maximum of 0.8 MDa (Toriyama and

Peters, 1980), and may be monocistronic mRNAs. Some of the very low molecular weight products could well have been a result of RNase in the preparations (Francki and Randles, 1973).

What does not emerge from any of these investigations with rhabdo- or reoviruses is how the genomic RNAs are replicated, nor can we exclude possible roles for host-specified components in the replicative function.

2.4.4. Host components in the replication of DNA viruses

Most research has concentrated on replication of the double-stranded genome of CaMV. Unlike positive-sense RNA viruses, there is clear evidence for involvement of host components at various stages of replication.

The DNA in the virus particle is curious in that it has three breaks, one in one strand and two in the other. In the infected cell, the DNA is uncoated, the discontinuities are joined, and the nucleic acid forms a minichromosome (Olszewski et al., 1982). This is a 'string of beads' structure in which the DNA is associated with groups of host proteins, with exposed linking sequences in between. Partial digestion with DNase followed by electrophoresis gave a 'ladder' pattern of CaMV DNA sequences, with a repeating unit of 195 bases between protein-containing beads. The minichromosome therefore resembles the nucleosome structure of the host chromatin. The CaMV minichromosome is assumed to be the transcriptional unit of the virus but not directly involved in replication (Menissier et al., 1983). The role of the host proteins is unclear, but may be similar to that in the structure of the host chromatin, where they maintain a condensed structure.

The minichromosomes are found in the nuclei of infected leaves, and this is where CaMV transcription is thought to occur (Guilfoyle, 1980; Ansa et al., 1982). Using nuclei isolated from CaMV-infected plants, Guilfoyle (1980) showed that the transcription of CaMV DNA had requirements for cations, and α-amanitin sensitivity, characteristic of the host DNA-dependent RNA polymerase II. Cooke et al. (1981) studied complexes formed by wheat-germ RNA polymerases I,

II and III with CaMV DNA, and showed that polymerase II formed the longest products in vitro. Guilfoyle and Olszewski (1983) reported that the sub-unit structure of host DNA-dependent RNA polymerase II was unchanged after infection.

The main transcripts from CaMV DNA found in infected plants are one of 2.3 kilobases, which codes for the 66 kDa inclusion body protein, and a transcript of 8 kilobases which is a full copy of the α-DNA strand (Covey and Hull, 1981). This RNA is thought to act as the template for synthesis of CaMV DNA by a reverse transcriptase activity. The experimental evidence, and possible mechanisms, have been reviewed by Hull and Covey (1983) and Hohn et al. (1983). It is thought that the synthesis of the α-strand DNA occurs using host methionine tRNA as a primer in strand initiation (Guilley et al., 1983). The proposed reverse transcriptase activity is currently being pursued, and is thought to be a virus-coded enzyme (Volovitch et al., 1984).

2.4.5. Host components in replication of viroids

Although viroids are the simplest known pathogens of plants, current knowledge suggests that they exist in diverse molecular forms: circular and linear RNAs, and multimers of the linear forms, of '+' and '-' senses (Branch and Robertson, 1984; the designations are arbitrary, because neither strand is thought to be translated). Therefore, replication could require a number of different enzyme activities. These include RNA polymerase(s), site-specific cleavage enzymes, and a ligase to form circular molecules from linear ones. Early evidence suggesting that viroid replication involved a DNA intermediate (Semancik and Geelen, 1975) has now been discounted (Zaitlin et al., 1980). But double stranded cDNA clones of HSV, for example, have been shown to be infectious, and cucumber plants inoculated with them produced authentic HSV RNA (Meshi et al., 1984). So while a DNA intermediate may not be part of the natural replicative cycle of a viroid, plants are nevertheless capable of dealing with one.

Viroid RNAs have no template activity in in vitro protein

synthesizing systems (reviewed by Sänger, 1982). Known sequences of the plus and minus forms have been collated by Symons et al. (1985) for a total of 8 viroids, and the potential viroid-coded polypeptides identified as sequences between initiation codons, and in-phase termination codons. Although these calculations show that some viroid-coded polypeptides are possible, ranging in size from a few amino acid residues up to about 100, there is no consistent pattern, and there is no evidence yet that they exist in vivo. Circular forms of RNA do not attach to eukaryotic ribosomes (Kozak, 1979); therefore only linear forms of viroids might serve as messengers. Thus the possibility that a short viroid-coded protein might modify a host replicase function cannot be ruled out (Sänger, 1984), but the evidence suggests strongly that viroid replication is entirely dependent on host components.

Investigations into which RNA polymerase activities might be involved in viroid replication have involved partially purified enzymes, isolated nuclei, and work at the level of whole cells using more or less specific inhibitors such as α-amanitin. The in vitro experiments have provided an embarrassment of riches, in terms of systems in which viroid RNAs can be transcribed. These include bacterial and bacteriophage enzymes (reviewed by Symons et al., 1985) as well as plant DNA-dependent and RNA-dependent RNA polymerases (Boege et al., 1982). Clearly the viroid RNA is a highly effective template, but the wide range of enzymes for which it is acceptable indicates that experiments in vitro may be of limited value in predicting how replication occurs in the plant.

The strongest evidence suggests a central role for DNA-dependent RNA polymerase II. This synthesized full-length copies of PSTV in vitro (Rackwitz et al., 1981). Viroid RNA was the most efficiently transcribed RNA species, and circular or linear forms were acceptable. The binding of this enzyme to viroid RNA has been studied in detail by Goodman et al., (1984). Their data indicated either one or two enzyme molecules bound per molecule of template.

Mühlbach and Sänger (1979) measured viroid RNA synthesis in PSTV-infected tomato protoplasts, and showed inhibition by concentrations of α-amanitin sufficient to inhibit polymerase II, but inactive against polymerase I. In a system using nuclei isolated from CEV-infected <u>Gynura</u>, Flores and Semancik (1982) found synthesis of linear and circular viroid RNAs. Actinomycin-D did not inhibit synthesis, but α-amanitin was strongly inhibitory. Sänger (1984) has discussed the possibility that DNA-dependent RNA polymerase III might also be involved.

Exclusion of DNA-dependent RNA polymerase I is entirely based on the α-amanitin evidence and on the insensitivity of viroid replication to actinomycin-D. This selectively inhibits polymerase I-dependent ribosomal RNA synthesis in plants (Fraser, 1975). However, both arguments would be negated if there were some other α-amanitin-sensitive step in viroid replication.

Various models have been proposed for viroid replication (Branch and Robertson, 1984; Symons <u>et</u> <u>al</u>., 1985), involving variations on the 'rolling circle' model of Brown and Martin (1965) for viral RNA synthesis. These models propose generation of multimeric copies by continuous transcription of a circular template, cleavage at specific sites to monomeric forms, and ligation of these to the circular form. The proposed intermediates have all been demonstrated in infected tissues, and enzymes capable of ligating linear viroid RNAs to form circular molecules have been found in wheat-germ (Branch <u>et</u> <u>al</u>., 1982). Cleavage does not seem to have been studied in detail, but it is likely that the site specificity is controlled by the viroid sequence rather than purely by the enzyme.

Branch and Robertson (1984) and Symons <u>et</u> <u>al</u>. (1985) have suggested that replication of the other sub-viral pathogens of plants, virusoids and some satellites, might involve mechanisms similar to those of viroid replication.

2.4.6. Is transcription of host genes after infection required for virus multiplication?

Expression of particular host genes might be required for various stages of the virus replicative cycle, and need not be confined to an involvement in transcription of viral RNA. Many workers have shown

that actinomycin-D will inhibit virus replication. This is taken to
indicate an inhibition of host DNA transcription necessary for virus
replication, and not a direct effect on transcription of viral RNAs.
Indeed, actinomycin-D has frequently been included in in vitro assays
of viral 'replicase' activities to inhibit traces of contaminating
host DNA-dependent RNA polymerase activities. Furthermore, virus
multiplication generally becomes insensitive to actinomycin-D at a
fairly early stage after inoculation, for example with TMV (Dawson
and Schlegel, 1976), CPMV (Rottier et al., 1979) or AMV (Nassuth et
al., 1983a). This suggests that the actinomycin-D sensitive stage is
an early event after inoculation, and that actinomycin-D does not
inhibit transcription of viral RNA templates directly. However, De
Varennes et al. (1985) pointed out that the apparent late
insensitivity to actinomycin-D might also be a result of failure of
inhibitor uptake later in the time-course, especially in experiments
with protoplasts, which develop cell walls during culture.

Although early treatment with actinomycin-D has been found to
inhibit multiplication of a number of viruses, there are also many
contradictory reports in the literature suggesting no inhibition for
the same viruses, for example TMV (Takebe and Otsuki, 1969) and CCMV
(Bancroft et al., 1975); further examples are given by Rottier et al.
(1979) and Mayo and Barker (1983). What emerges is that the effect
might be dependent on experimental differences, such as whether
protoplasts or whole leaves were used, or possibly trivial
explanations such as different rates of actinomycin-D uptake.
However, it remains likely that the effects of actinomycin-D on plant
virus multiplication could be various and complex. Lockhart and
Semancik (1969) found that the inhibitory effect of actinomycin-D
depended on the host in which a single virus was multiplied.

Turner and Dawson (1984) studied the early, actinomycin-D sensitive
step during infection of cowpeas with CCMV, CPMV and the cowpea
strain of TMV. Leaves were inoculated with one virus and incubated
until its multiplication had reached the actinomycin-D insensitive
stage. The leaf was then inoculated with a second virus, and
actinomycin-D was applied at various times. For all combinations of

viruses tested, inhibition of the second virus by actinomycin-D was not prevented by the successful infection by the first virus. This may suggest that the presumed host transcription which was inhibited by actinomycin-D early after infection is virus-specific. Alternatively, the host-coded function might be non-specific but unstable, and required only for the early hours of infection.

Some authors have attempted to discover which stage of virus replication might be inhibited by actinomycin-D treatment. In CPMV-infected cowpea protoplasts, Rottier _et al_. (1979) found that synthesis of the two virus coat proteins continued and empty capsids were produced, under conditions where virus multiplication was completely inhibited. They also demonstrated synthesis of a further four virus-coded proteins. But De Varennes _et al_. (1985) suggested that the apparent insensitivity of viral protein synthesis to actinomycin-D might be partly indirect, and a result of the known ability of the inhibitor to slow down degradation of messenger RNA.

Nassuth _et al_. (1983a) found that AMV coat protein synthesis was less inhibited than synthesis of (+) and (-) strand viral RNAs, in actinomycin-D treated cowpea protoplasts. These results may suggest that production of sub-genomic messenger RNAs is less sensitive to actinomycin-D than production of infectious, genomic RNAs.

Mayo and Barker (1983) did not find a differential effect of actinomycin-D on TRV RNA and coat protein synthesis in tobacco protoplasts. Actinomycin-D did not inhibit TRV or TMV infection when added to protoplasts immediately after inoculation with viral RNAs, but was inhibitory for a few hours after inoculation with virus particles. They suggested that host transcription might be required for the uncoating process, but also considered an alternative explanation, in which uncoating merely provided a delay period, during which actinomycin-D was able to inhibit host transcription required for some other early viral function. They also pointed out that an involvement in uncoating would be unlikely to explain the AMV and CPMV results discussed above.

It should be noted that in a few special cases, actinomycin-D has been reported to increase virus multiplication. These are where the inhibitor has prevented host transcription leading to activation of a genetically-controlled resistance mechanism, and are considered in that light in Chapter 6. On balance, the experiments with actinomycin-D and other inhibitors of host RNA synthesis in virus-susceptible plants suggest that transcription of host genes at an early stage of infection is required for virus multiplication. It is not yet known how many host genes are involved and how the proteins they specify might be involved in replication.

2.5. ASSEMBLY AND AFTER

2.5.1. Other components of the virion provided by the host

Many plant viruses are known to contain bivalant cations which are important in stabilizing the particle. Ca^{2+} and Mg^{2+} are typically present in amounts close to 1 cation per protein subunit (Hsu et al., 1976; Hull, 1978). These cations are undoubtedly supplied by the host systems for uptake and translocation of mineral nutrients. It is not known whether virus infection induces any special patterns of mineral nutrition related to virus assembly, although changes in the mineral composition of virus-infected plants have been found (Bergman and Boyle, 1962; Chandra and Mondy, 1981).

The particles of some viruses are also stabilized by polyamines, which bridge negatively-charged phosphate residues in the nucleic acids. TYMV has been most studied, and has been shown to contain 200–700 molecules of spermidine per virion, and a smaller amount of spermine (Cohen and Greenberg, 1981). The polyamines are of host origin. Balint and Cohen (1985) showed that TYMV-infection of Chinese cabbage protoplasts induced an accumulation of spermidine. The implication of radiolabelling experiments with methionine, a precursor of spermidine, was that newly synthesized polyamines were incorporated into virus particles in preference to those from the existing pool. However, inhibition of spermidine synthesis led to increased utilization of pre-existing spermidine, and increased incorporation of spermine into the newly formed virions.

Plant rhabdoviruses and TSWV are unique in that their particles are

Fig. 2.5. Electron micrograph showing assembly of broccoli necrotic
yellows virus (Rhabdovirus). Virus particles (P) of different
lengths are enclosed in a membrane (M) of host origin, and attached
to a membraneous structure by short stalks (S). The preparation was
negatively-stained with methylamine tungstate. The bar shows 200
nm. Micrograph by Colin Clay, National Vegetable Research Station.

enclosed by a lipoprotein membrane, which is derived from the host.
For rhabdoviruses, current thinking is that the membrane is derived
either from pre-existing host membranes, such as the nuclear membrane
or endoplasmic reticulum, or from cytoplasmic viroplasms induced by
the virus (Francki et al., 1981). Different rhabdoviruses acquire
their membranes from characteristic sites. Fig. 2.5 shows particles
of the rhabdovirus BNYV of different lengths apparently growing out
of a membraneous complex. The interpretation is that assembly of the
inner components of the virion is occurring in membraneous extensions
from the matrix.

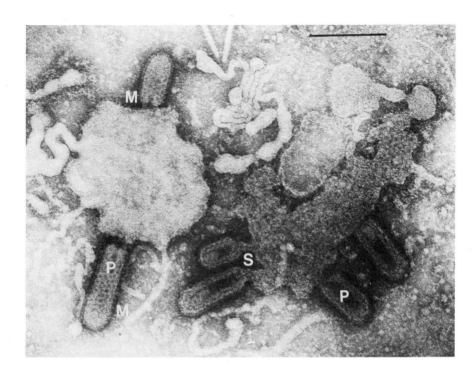

2.5.2. Possible involvement of host components in assembly

It has been shown for a number of types of plant viruses in both rod-shaped and spherical groups that assembly from the nucleic acid and protein components can be accomplished in vitro, given suitable conditions of pH, salt and temperature (e.g. Verduin et al., 1984; Butler, 1984). Furthermore, hybrid particles can be reconstituted in vitro between the nucleic acids and proteins of quite unrelated viruses. These results might be taken to indicate that virus assembly does not require the participation of host-specified components. However, they do not exclude the possibility that host components or structures might be important in assembly in vivo, for example by promoting compartmentalization of virus-specified components, or providing transport systems between sites of synthesis and sites of assembly.

TYMV is probably the best understood case where the host appears to provide a known location for assembly with possible advantage to the virus. Ushiyama and Matthews (1970) and Lafleche and Bove (1971) showed that after TYMV infection, peripheral vesicles formed on the chloroplasts, and suggested that they might be involved in virus replication. Matthews (1973) reviewed evidence showing that these vesicles are connected to the cytoplasm by a 'neck', and discussed the relationship of the vesicle membranes to the chloroplast membranes. The replicase and template complex of the virus is thought to be associated with the inner membrane of the vesicle, and assembly to occur at the neck where coat protein arrives after synthesis on cytoplasmic ribosomes.

CaMV assembly occurs in characteristic inclusion bodies (Shepherd et al., 1980). The matrix protein of these bodies is virus coded (Covey and Hull, 1981), as is a protein specifying aphid transmissibility (Woolston et al., 1983). Shockey et al. (1980) found a number of other proteins in the inclusion bodies, but there was no evidence that these were host coded. Instead, it was likely that they were capsid protein and its degradation products. However, their procedure for purification of inclusion bodies involved the use of non-ionic detergents, and it remains possible that host membranes were removed during extraction. Whether host components contribute to the activity of inclusion bodies in assembly remains much in doubt.

The characteristic inclusion bodies associated with potyvirus infections have not been directly associated with assembly or any other function in virus multiplication (Hollings and Brunt, 1981). The major proteins of both cytoplasmic and nuclear inclusion bodies appear to be virus-specified (Dougherty and Hiebert, 1980).

2.5.3. Virus transport within the plant

From the initially-infected cell, the virus moves from cell to cell, may enter the phloem and be transported over long distances, and may be unloaded from the phloem to initiate systemic infection. A few viruses also appear to be transported in the xylem (reviewed in Atabekov and Dorokhov, 1984). Clearly, these different stages may involve different processes, and might involve active or passive transport mechanisms. For each type of transport, one can ask what form of virus material is moved; whether there are virus-specified components with a transport function, and whether there are host-specified components. In the last case, the nature of the interaction between the host and virus components is also of interest. At present, little is known about the virus-specified aspects, and almost nothing about host-specified transport functions.

Atabekov and Dorokhov (1984) reviewed evidence that infection passes from cell to cell through the plasmodesmata, and that viruses may modify the structure of plasmodesmata. But nothing is known of the mechanisms. For TMV, they also reviewed evidence that the infectious form transported is not the entire virion, but a virus ribonucleoprotein particle (vRNP) or 'informosome'. This contains RNA plus a number of proteins, which appear to be virus-specified.

A TMV mutant, Ls1, is temperature sensitive in cell-to-cell spread (Nishiguchi et al., 1980), and this is associated with a single amino acid substitution in the 30 kDa protein (Ohno et al., 1983). It is not clear how the altered protein may operate. Shalla et al. (1982) suggested that there were significantly fewer plasmodesmata between cells of leaves infected with Ls1 and held at the restrictive temperature, than between cells of similarly infected leaves either held at the permissive temperature, or inoculated with the parent wild type strain at either temperature. However, the differences in actual numbers of plasmodesmata did not seem to be large enough or consistent enough to explain the failure of spread of Ls1.

Functions involved in transport are specified by the M-component RNA of CPMV. In its absence, the B-component RNA was shown to replicate in primarily inoculated cells, but not to spread from cell to cell (Rezelman et al., 1982).

An interesting result on possible virus-specified transport functions was obtained by Taliansky et al. (1982). They showed that systemic PVX infection of tomato plants containing the Tm-2 gene for TMV resistance could allow TMV to spread systemically. In plants inoculated with TMV alone, the virus was strongly localized by the resistance gene. They suggested that the PVX transport function may have modified the host to permit transport of viruses, while TMV could not do so. It would be interesting in this context to examine virulent isolates of TMV which overcome the Tm-2 gene, to see if they differ from avirulent isolates in their 30 kDa proteins. Other aspects of this complementation experiment are considered in the context of resistance mechanisms in Chapter 6.

Viruses are able to load into, and unload from the phloem, but nothing is known of how they might exploit host mechanisms in so doing. Long-distance transport within the phloem is thought to be passive, with the virus being carried in the assimilate flow. Helms and Wardlaw (1978) showed that the rate of translocation of TMV in the phloem of Nicotiana glutinosa plants was much less than that of photosynthetic assimilate. They attributed this to impedances to movement of the large TMV particles by protein filaments and sieve plates, as well as to uncertainties of measurement associated especially with phloem loading. Gill and Chong (1981) showed that infection of oats with various isolates and mixtures of BYDV caused changes in the ultrastructure of phloem cells which were virus strain-related. However, there was no clear link with host-specified processes involved in virus transport. Finally, De Zoeten and Gaard

(1983) suggested that transport of PEMV from the infection sites to phloem elements occurred in vesicles which were presumably of host origin.

2.5.4. Participation of the host in virus transmission

Viruses are certainly able to invade those parts of the plant which are disseminated: the pollen and the seed. Both routes can aid virus transmission (e.g. Couch, 1955; Frosheiser, 1974). There appear to be two routes of transmission: externally on the seed coat or pollen grain wall, and within the male gametophyte or embryo (reviewed in Matthews, 1981). External transport is obviously purely passive in biochemical terms, but for both routes of internal transport, there are possibilities of interesting biochemical interactions between plant and virus. Thus the susceptibility of lettuce varieties to seed transmission of LMV varies (Couch, 1955), and resistance to seed transmission of BSMV in barley is controlled by a single recessive gene (Carroll et al., 1979). Virus-specified factors may also control the extent of seed transmission (e.g. Hanada and Harrison, 1977). The biochemical basis of these differences remains to be established, but an understanding of how some varieties may prevent invasion of the embryonic tissue by viruses could assist in development of methods to prevent seed transmission.

2.6. CONCLUSION

There is good understanding of how some host functions are involved in virus replication; other aspects are less clear. The best and most detailed evidence for host involvement is in the 'general services' area: provision of energy, precursors and translational machinery. More research is required to understand how the host may be involved in more specialized aspects of virus multiplication, such as nucleic acid replication and in transport. Progress in these areas also requires a fuller understanding of how virus-coded functions operate.

CHAPTER 3

Effects of Virus Infection on Host Nucleic Acid Metabolism

3.1. DNA

3.1.1. DNA synthesis

There have been comparatively few studies of the effects of viruses on host DNA synthesis and properties. Fraser (1972) showed that accumulation of DNA in tobacco leaves was inhibited if the leaves became systemically infected by TMV at an early stage of development. The amount of inhibition was about as much as the general inhibition of leaf growth measured as length or weight. Atchison (1973) found a drop in the rate of DNA synthesis at the time of invasion of Phaseolus vulgaris root tips by TRSV. This was followed by a fall in the mitotic index to half that in healthy root tips, then by recovery to the normal level. In contrast with these reports, Gense (1980a) found that inoculation of tobacco leaves with TMV giving a hypersensitive or systemic response, or physical wounding, caused a stimulation of DNA synthesis. This was nuclear but non-mitotic; its significance is unknown (Gense, 1980b).

3.1.2. Cytogenetic effects

Several cytogenetic effects of viruses have been reported, but the mechanisms are incompletely understood. BSMV or WSMV infection of maize can cause the 'aberrant ratio' (AR) phenomenon, whereby inheritance in the F_2 and back-cross generations at certain loci deviates consistently from Mendelian expectation, although other loci on the same chromosome segregate normally (Sprague and McKinney, 1966; 1971). Brakke (1984) considered several possible mechanisms for AR, and also discussed evidence suggesting that viral infection might be mutagenic. The evidence did not permit firm conclusions on either effect, but an intriguing possibility was that the virus might serve as a vector for a host controlling element or regulatory RNA. This might be transmitted with the virus as a pseudovirion, and might depend for its replication on the simultaneous replication of the virus in the host.

Mottinger and Dellaporta (1983) and Dellaporta et al. (1983) reported stable and unstable mutations at various loci, associated with BSMV-infection of maize. There was no evidence for insertion of viral sequences into the host DNA. They suggested that maize transposable elements (McClintock, 1978) may have been activated as a result of the stress of virus infection.

Viruses and virus-like agents have also been implicated in the cytoplasmic male-sterility trait of a number of host species (Grill, 1983; Grill et al., 1983; Sisco et al., 1984), and possibly they interact in some way with the mitochondrial genome. Sandfaer (1979) found that strains of BSMV differed in their ability to cause sterile flowers, but that all strains examined increased the frequency of triploid and aneuploid seeds in wheat and barley.

3.1.3. Effects on chromatin

Van Telgen et al. (1985b; 1985c) extracted chromatin from TMV-infected tobacco leaves. They showed that a protein with an apparent molecular weight of 20 kDa could be dissociated from the chromatin in 6M urea, and identified it serologically as the coat protein. Another protein with an apparent molecular weight of 116 kDa remained associated with the chromatin under these conditions, but was dissociated in 0.5 M NaCl. Subsequently, Van Telgen et al. (1985a) showed that this protein was very similar to or identical with the virus-coded 126 kDa protein (Scalla et al., 1978; Goelet et al., 1982) (Fig. 3.1).

The significance of the association of this protein with chromatin

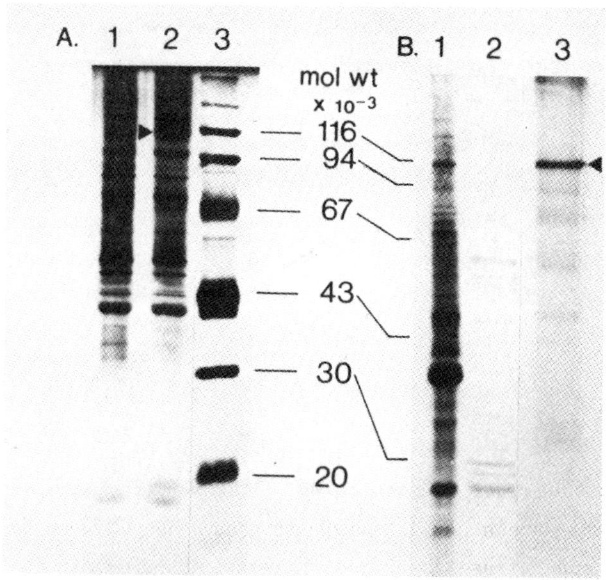

Fig. 3.1. Electrophoretic patterns on polyacrylamide gels, of protein fractions extracted from tobacco leaves. (A), Nuclear proteins extracted from purified nuclei. Track 1, from healthy plants. Track 2, from TMV-infected plants. The arrow indicates the position of the '116 kDa' polypeptide. Track 3, marker proteins. (B), Track 1, chromatin-associated proteins from mosaic-diseased leaves. Track 2, products obtained after translation of TMV RNA in a rabbit reticulocyte *in vitro* system, stained with Coomassie blue. Track 3, autoradiogram of track 2, showing detection of 35-S methionine-labelled translation products. The arrow indicates the position of the 126 kDa polypeptide. From Van Telgen *et al*. (1985a), by permission of the authors and Academic Press.

is unknown. Van Telgen *et al*. (1985a; 1985b) suggested that it might be involved in control of host gene expression during development of visible symptoms. The concentration (measured as μg/unit volume) of the 116/126 kDa protein in the nuclear fraction was about eight times that in the cytoplasmic soluble fraction. However, measured as absolute amounts (μg/leaf), both soluble and membrane fractions contained about four times as much of the protein as the nuclear fraction, i.e. the bulk of this protein in the cell was located outside the nucleus. This may be relevant to the indirect evidence, which includes arguments drawing on base sequence homology between the TMV genome and the replicase gene of poliovirus (Kamer and Argos, 1984), that the 126 kDa protein or the readthrough product of the same sequence is involved in TMV replication in the cytoplasm or membranes. Such a role would not preclude a separate function involving association with chromatin.

Balazs *et al.* (1979) reported a change in histone-like proteins in tobacco reacting hypersensitively to TMV, but not in plants with a systemic infection. No information was given on whether these histone-like proteins were associated with DNA.

3.2. RNA

Early investigations concentrated on changes in the metabolism of total (host plus viral) or host RNA overall, and have been reviewed (Mundry, 1963; Fraser, 1972). Later work exploited the increasing ability to isolate RNAs with specific characteristics and functions. As most plant viruses are ribonucleoproteins, we can envisage several areas of interaction between virus multiplication and host RNA metabolism. Thus viral RNA and host RNA share the same ultimate precursor pools of nucleoside triphosphates, while the existing host RNA might be a potential source of precursors for further viral RNA synthesis. On the other hand, the virus has a requirement for host ribosomes and transfer RNAs for synthesis of viral protein. This might also involve an interaction with host messenger RNA metabolism.

Several questions may be asked about the interaction between host and virus:

a) descriptive aspects: what are the effects of infection on rates of synthesis, turnover and utilization of host RNAs?

b) how do these changes relate to virus multiplication - for example do they create circumstances which might favour multiplication?

c) what is the significance of the changes for the host, in terms of direct effects on growth and development?

d) where virus infection does cause changes in host RNA metabolism,

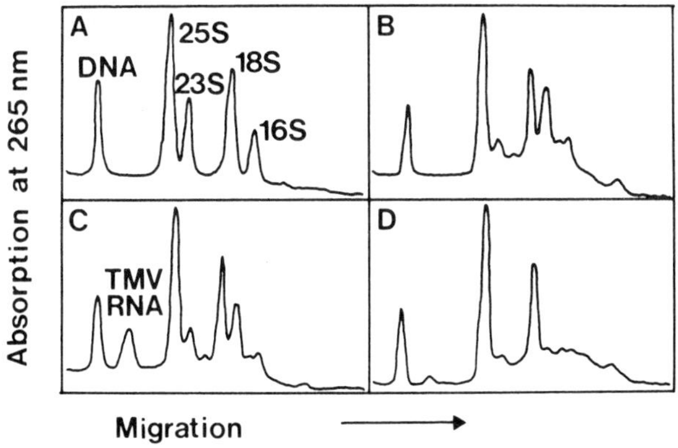

Fig. 3.2. Polyacrylamide gel electrophoresis of total nucleic acids extracted from tobacco leaves. (A) healthy leaf, 50 mm long.(B) healthy leaf, 170 mm long.(C) leaf of the same age as (B), 4 days after inoculation with TMV strain vulgare.(D) leaf of the same age as (B), 4 days after inoculation with TMV strain flavum. The numbers indicate the nominal sedimentation coefficients of the cytoplasmic (25S and 18S) and chloroplast (23S and 16S) ribosomal RNAs. Reproduced from Fraser (1969).

what biochemical control mechanisms are involved?

3.2.1. Ribosomal RNA (rRNA)

Virus-induced changes have been studied by measuring amounts of cytoplasmic (80S) or chloroplast (70S) ribosomes, and amounts or rates of synthesis of their rRNA components (25S plus 18S, and 23S plus 16S respectively). Effects on synthesis and stability of mitochondrial ribosomes do not appear to have received attention.

Chloroplast rRNA. Most studies have indicated more severe effects of infection on chloroplast ribosomes and rRNAs than on cytoplasmic components. In TMV-infected tobacco, Hirai and Wildman (1969) and Fraser (1969) found inhibition of incorporation of radioactive precursors into chloroplast rRNA while cytoplasmic rRNA synthesis was little affected. Fraser (1969) also noted accelerated degradation of chloroplast rRNA. Fig. 3.2 shows that in young, healthy tobacco leaves, the peak areas of the 23S and 18S chloroplast rRNAs were in a ratio of about 2:1. This, with the known two-fold difference in molecular weights, indicates that they were present in a molar ratio close to 1:1. As the leaf aged, the 23S was lost, and a series of smaller degradation products appeared. TMV infection accelerated this degradation, especially when the *flavum* or other strains with defective coat protein were used (Schuch, 1974). The interpretation is that some defective coat protein denatured during virus multiplication, and caused disruption of chloroplast structure, with consequent exposure of the chloroplast ribosomes to degradative nucleases (Jockusch and Jockusch, 1968). From Fig. 3.2, it is clear that cytoplasmic rRNAs were not degraded in this way.

Other authors who have reported selective degradation of chloroplast ribosomes include Randles and Coleman (1970) for LNYV-infected *Nicotiana glutinosa*; Mohammed and Randles (1972) for TSWV-infected tobacco, and White and Brakke (1982) for barley and wheat leaves infected with BSMV or WSMV. Magyarosy *et al*. (1973) found that SqMV-infection of squash plants increased the concentration of cytoplasmic ribosomes but decreased the concentration of chloroplast ribosomes. The latter effect appeared to be a result of a reduced number of chloroplasts rather than degradation of their ribosomes, as chloroplast ultrastructure was unaffected by the virus. Roberts and Wood (1981a) found a decrease in both cytoplasmic and chloroplast ribosomes in CMV-infected tobacco. It is interesting that a fungal infection has also been reported to cause preferential degradation of chloroplast ribosomes (Bennett and Scott, 1971).

The differential effects of viral infection on chloroplast and cytoplasmic ribosomal components have generally been measured in the earlier stages of pathogenesis. In the longer term, major effects on cytoplasmic rRNA synthesis, stability and turnover have certainly been observed, although they were less rapid than those on chloroplast rRNAs.

Cytoplasmic rRNA. In healthy tobacco leaves, the rate of rRNA synthesis increased as the leaf expanded, then declined from about

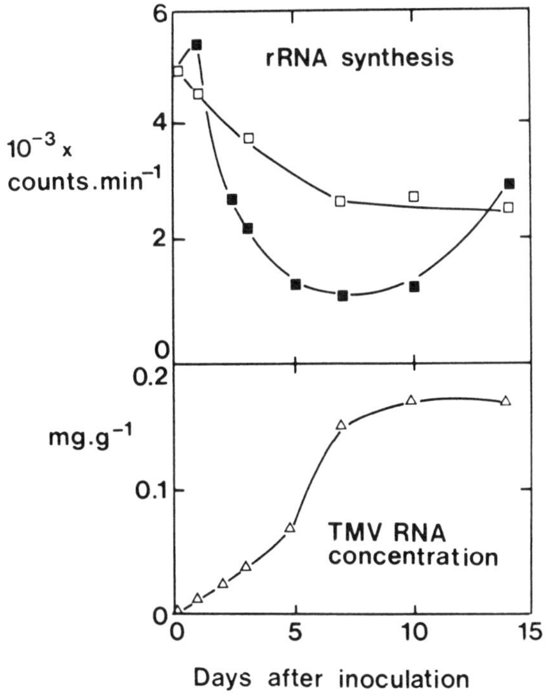

Fig. 3.3. Rates of synthesis of cytoplasmic ribosomal RNA and TMV RNA in tobacco leaves. Leaves were pulse-labelled for 5 hours with 32-P phosphate, and incorporation into specific RNAs measured after fractionation on polyacrylamide gels. Leaves were 170 mm long at day 0. (□) rRNA synthesis, healthy leaves; (■) rRNA synthesis in leaves inoculated with TMV strain vulgare at day 0. (△) Shows the accumulation of TMV RNA. Reproduced from Fraser (1973a).

the stage where full size was reached (Fraser, 1973a). TMV infection at a very early stage caused a lasting inhibition of the rate of cytoplasmic rRNA synthesis. Fig. 3.3 shows the effects of inoculating a leaf at the end of the expansion phase. After a brief period when rRNA synthesis was slightly stimulated by infection, it was lower than in healthy leaves, by as much as 60%, during the period of viral RNA accumulation. After TMV RNA accumulation had ceased, rRNA synthesis in infected leaves rose again to the healthy leaf level. This would be consistent with an inhibition of rRNA synthesis due to competition for precursors from TMV RNA synthesis.

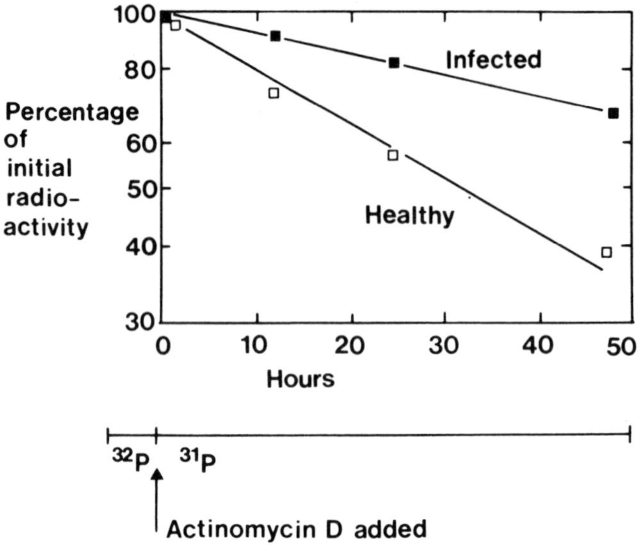

Fig. 3.4. Measurement of the rates of turnover of cytoplasmic ribosomal RNA in healthy (□) and TMV strain vulgare-infected (■) tobacco leaves. Leaves 130 mm long were inoculated two days before radioactive incubation for 4.5 hours with 32-P phosphate, infiltrated with 50 µg/ml actinomycin-D and transferred to non-radioactive medium. Radioactivity of cytoplasmic ribosomal RNA (25S plus 18S) was measured after fractionation on polyacrylamide gels. Reproduced from Fraser (1973a).

Kubo (1966) and Kubo and Tomaru (1968) have also reported stimulation of cytoplasmic rRNA synthesis shortly after infection, but inhibition during the period of most active TMV RNA accumulation. Pring (1971) found an inhibition of host rRNA synthesis during multiplication of BSMV in barley.

In infected tomato roots, where the total amount of TMV RNA accumulated was much lower than in tobacco leaves, infection caused a major stimulation of the rate of cytoplasmic rRNA synthesis (Fraser et al., 1973).

When strips from healthy or TMV-infected leaves were pulse-labelled with ^{32}P, then transferred to non-radioactive phosphate with a high concentration of actinomycin-D to prevent further rRNA synthesis, the

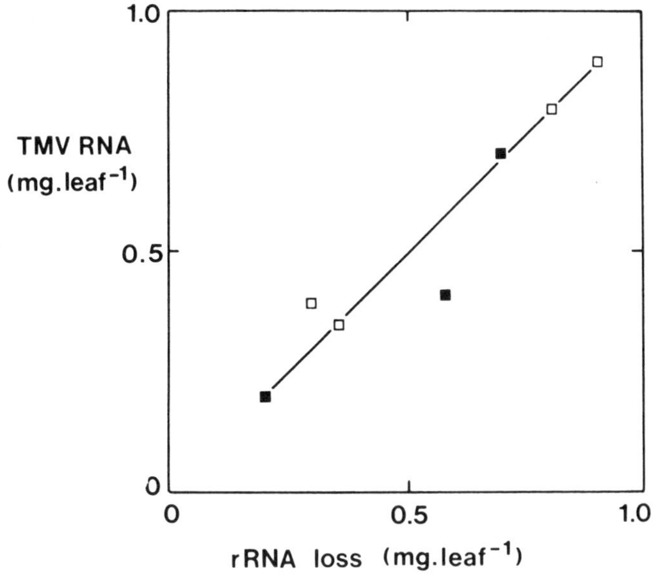

Fig. 3.5. Comparison of the amounts of TMV RNA accumulated in tobacco leaves inoculated at different ages, with the amount of loss of cytoplasmic ribosomal RNA content during the period of virus multiplication. (□) strain *vulgare*; (■) strain *flavum*.

decline in radioactivity of the rRNA could be taken as a measure of the rate of turnover. The data in Fig. 3.4 suggest that in mature, healthy leaves, the rRNA half-life was 32 hours, whereas in comparable infected leaves, it was 87 hours. In this case, therefore, TMV infection actually decreased the turnover of cytoplasmic rRNA, in contrast to its degradative effects on choroplast rRNA.

The effect on cytoplasmic rRNA stability was confirmed by examination of changes in the rRNA content of healthy leaves after infection (Fraser, 1972). Expanded, healthy leaves suffer a net loss of cytoplasmic rRNA as they age. Infected leaves showed a slower rate of loss. Fig. 3.5 relates the amounts of TMV RNA accumulated in leaves inoculated at different developmental stages, with the amount of cytoplasmic rRNA degraded during virus multiplication. There was clearly a strong correlation. This suggests that in mature leaves, turnover of cytoplasmic rRNA serves as the major source of precursors for TMV RNA synthesis, and that the amount of turnover may limit the amount of viral RNA made.

An early report (Reddi, 1963) proposed that TMV infection actually accelerated the breakdown of host cytoplasmic rRNA, and that the degradation products were re-incorporated into viral RNA. This is in contrast to the retardation of degradation suggested here. However, examination of the experimental design in Reddi's work shows that leaves were inoculated at different stages of development, and that what was measured was inhibition of rRNA accumulation, rather than accelerated breakdown. Using a different experimental design, Babos (1966) also found no evidence for any acceleration of host cytoplasmic rRNA turnover after TMV infection.

Possible consequences and mechanisms. The various effects of TMV on rRNA metabolism may be interpreted as producing circumstances which favour virus multiplication. Chloroplast ribosomes, as far as is known, play no part in TMV protein synthesis (Paterson and Knight, 1975; see Chapter 2.3.1). Eliminating their synthesis and increasing turnover of chloroplast rRNA early in infection would reduce competition and increase the supply of nucleoside triphosphate precursors. Cytoplasmic ribosomes, on the other hand, are required for virus protein synthesis, and their increased stability after infection of mature leaves would serve, if required, to maintain a high concentration for this purpose.

It is not known how the effects on synthesis are mediated, nor, with the possible exception of the mechanism involving disruption of chloroplasts by denatured virus coat protein, how virus may affect turnover. Hirai and Wildman (1969) did show that the effects of TMV on chloroplast rRNA synthesis were detectable when synthesis was measured in chloroplasts isolated from infected leaves. Roberts and Wood (1981a) found a loss of chloroplast DNA-dependent RNA polymerase activity after infection. These results would suggest a direct control mechanism acting within the chloroplast.

Whenham and Fraser (1981) showed that TMV infection of tobacco increased leaf concentration of the plant growth regulator abscisic acid (ABA), and that treatment of leaves with similar amounts of ABA could stimulate host and viral RNA syntheses (Whenham and Fraser, 1980). This might have contributed to the slight stimulation of cytoplasmic rRNA synthesis observed after inoculation, before a separate competitive inhibition from viral RNA synthesis became predominant. Furthermore, while there is no direct explanation for the delayed loss of cytoplasmic rRNA from infected, mature leaves, ABA has been shown to inhibit some aspects of senescence on intact plants (Hall and McWha, 1981).

The majority of investigations of ribosomal RNA metabolism in infected leaves have involved tobacco, and most have used TMV. This virus multiplies to a very high level (see Fig. 1.1) and its replication clearly makes severe 'economic' demands on the host. Other viruses which multiply much less, but which nevertheless may cause equally or more severe visible symptoms, might have quite different effects on host ribosomal RNA metabolism, and deserve further study.

3.2.2. Transfer RNA (tRNA)

Effects of virus infection on transfer RNA metabolism do not seem to have been studied in detail. Fraser (1972) found that when tobacco leaves were inoculated with TMV at early or late stages of development, the effects on tRNA content were similar to those on cytoplasmic rRNA content.

3.2.3. Host messenger RNA (mRNA)

During virus multiplication, synthesis of viral coat protein may take up the major part of the host capacity for total protein synthesis, as discussed in Chapter 4. This clearly raises questions about how host protein synthesis is controlled. At the mRNA level, virus infection might reduce host protein synthesis by decreasing the rate of host mRNA synthesis, by increasing its turnover, or by reducing translation. The first two mechanisms would cause a reduction in total host mRNA content, while the third would not necessarily do so.

Host mRNA metabolism can be studied by several methods, which give different types of information. A few authors have studied polyribosome profiles, although results here may be influenced by association of viral RNAs with the ribosomes (McCarthy et al., 1970). Randles and Coleman (1972) reported a decline in the proportion of ribosomes occurring as polyribosomes during development of LNYV symptoms on Nicotiana glutinosa, and suggested that this was the cause of an observed reduction in the overall rate of protein synthesis. Bol et al. (1976) found that almost all of the rapidly-labelled RNAs associated with polyribosomes of AMV-infected tobacco were viral, suggesting an almost complete exclusion of host mRNAs from the translational process.

In vitro translation of total or sub-fractionated mRNA preparations from infected leaves could potentially give very specific information about the changes in amounts of mRNAs coding for specific host, as well as viral polypeptides. Carr (1985) compared electrophoretic separations of proteins radiolabelled in vivo, and by in vitro translation of polyadenylated mRNA (poly(A)mRNA) extracted from leaves. He compared healthy Xanthi-nc tobacco leaves with those reacting hypersensitively to TMV infection. Whereas the pattern of proteins synthesized in vitro appeared to be unchanged by infection, there were differences in the proteins labelled in vivo. Some disappeared and some appeared after infection. Carr reached the preliminary conclusion that the poly(A)mRNA populations of healthy and infected leaves could be similar, but that controls at the translational level might influence the pattern of proteins synthesized in vivo.

A particularly interesting finding concerns the 'pathogenesis-related' proteins, which are not normally detectable in healthy leaves. Messenger RNAs for these proteins were found to be present, and could be translated in vitro, in poly(A)mRNA preparations from healthy leaves (Carr et al., 1982). But Hooft van Huijsduijnen et al. (1985), using cDNA probes to the PR1 group of pathogenesis-related proteins, were able to detect only very low levels of PR1 mRNA in total RNA prepared from healthy leaves. This is discussed further in Chapter 4.

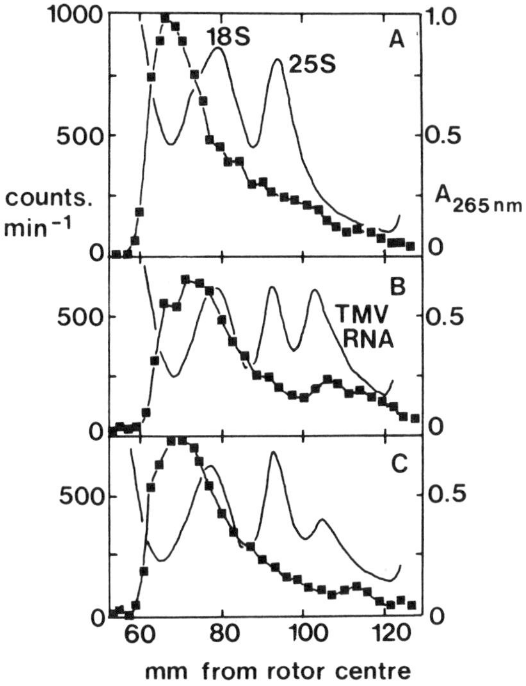

Fig. 3.6. Fractionation by sucrose density gradient centrifugation of nucleic acids from tobacco leaves. (A), healthy; (B), infected with the common strain of TMV, and (C), infected with the _flavum_ strain. The continuous line shows absorbance at 265 nm. (■) indicates the distribution of host poly(A)mRNA, measured by hybridization of 3-H polyuridylic acid to gradient fractions. From Fraser and Gerwitz (1980).

Although some viruses have poly(A) sequences in their genomic or messenger RNAs (see, for example, Davies _et al._, 1979; Milner and Jackson, 1983), others such as TMV do not (Fraser, 1973b; Siegel _et al._, 1976). Under the latter circumstances, the concentration of that fraction of host mRNA which is polyadenylated (Gray and Cashmore, 1976) can be studied by hybridization with a radioactive complementary homopolynucleotide such as ^{3}H-polyuridylic acid. Fraser and Gerwitz (1980) showed that in TMV-infected tobacco leaves, the ratio of poly(A)mRNA to total (ribosomal plus transfer) RNA was the same as in comparable healthy leaves throughout the period of virus multiplication. Virus infection also did not alter the size distribution of host poly(A)mRNAs (Fig. 3.6). Host protein synthesis, on the other hand, was inhibited by 50 - 75% during virus multiplication (see Chapter 4.2). These results therefore suggest that TMV infection had no effect on host poly(A)mRNA synthesis or turnover, and that control of host protein synthesis must have been at the translational level. This could have been a consequence of the high competitive efficiency of the TMV coat protein mRNA in translation.

In contrast to the limited data on host mRNA concentration after virus infection, studies of changes after fungal infection generally have shown severe reduction in mRNA concentrations (Simpson et al., 1979; Manners and Scott, 1985).

3.3. CONCLUSION

It is clear that current knowledge of changes in host nucleic acid metabolism is greatest for those types of nucleic acid where synthesis and turnover can be most easily investigated. The comparatively large amount of information on ribosomal RNA, albeit from a very restricted range of host and virus combinations, can be contrasted with the limited information on possible effects of viruses on expression of individual host genes. Yet the most interesting questions are how expression of individual genes might be controlled, and on how they might be involved in processes such as assisting virus multiplication, or in host resistance. Current techniques which allow preparation of specific cDNA probes, and the ability these give to isolate genes from genomic libraries and also to assay for the gene products, should be invaluable in characterizing changes in gene expression at several levels.

CHAPTER 4

Virus Infection and Host Protein Metabolism

4.1. DIFFERENT APPROACHES TO HOST PROTEIN METABOLISM

The replicative cycle of a virus may interact with the protein metabolism of its host at several levels. As considered in Chapter 2, certain host proteins might be directly involved in the biochemical processes of viral nucleic acid and protein synthesis. Other host proteins may be involved in recognition events during the development of pathogenesis or expression of resistance mechanisms, although the evidence for these is still tenuous.

Virus infection might also have a number of indirect effects on host protein metabolism. These could be related to the production of symptoms, to the possible competitive inhibition of plant growth by synthesis of significant amounts of a 'foreign' protein, and to possible viral controls on expression of the host genome.

Research on changes in host protein metabolism induced by viral infection has been largely descriptive, but more recent approaches have tried to interpret the changes in terms of control of the processes of pathogenesis. Investigations may be classified into five main groups:

a) Changes in the amounts, synthesis and turnover of host protein as a whole, and their relationship to the accumulation of viral protein;

b) Fractionation of total or defined fractions of host protein, for example by polyacrylamide gel electrophoresis, to allow examination of changes in individual bands and assessment of the overall effects of infection;

c) Studies of enzymes with known activities which can be assayed, or of proteins with known structural or catalytic roles extracted from purified organelles such as the chloroplasts;

d) Detailed studies of proteins with defined properties (although sometimes of unknown function) which can readily be isolated;

e) Studies of the metabolism of amino acids as precursors or breakdown products.

Some examples of each of these types of study will now be considered; inevitably the divisions are somewhat arbitrary and there are some overlaps.

4.2. GROSS CHANGES IN HOST PROTEIN METABOLISM

Some viruses, such as TMV, multiply so successfully that their coat protein becomes by far the commonest single protein in the leaf, and represents the majority of the protein content by weight (see Fig. 1.2). This 'foreign' protein metabolism is likely to place considerable stress on host protein synthesis and turnover, and may also affect processes involved in plant growth and development.

In contrast, many viruses accumulate to limited extents, and the viral proteins represent only a small fraction of total leaf protein. Most potyviruses fall into this category. There is little possibility of overall competitive effects on host protein metabolism in these cases, but other levels of interaction are possible. Questions which can be raised include:

a) Does accumulation of large amounts of viral protein occur at the expense of host protein synthesis, or by accelerated turnover of host protein?

b) Is overall host capacity for protein synthesis diminished by non-competitive, pathological effects of the virus?

c) Do the observed changes in host protein metabolism induced by the virus explain some or all of the changes in growth and development? (This question is also addressed in Chapter 7).

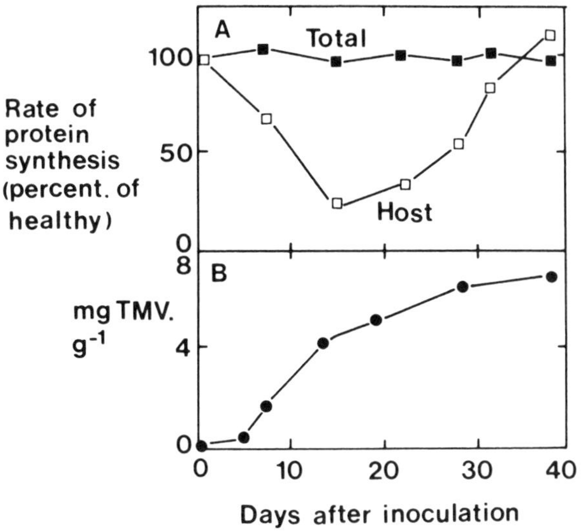

Fig. 4.1. Changes in rates of protein synthesis during TMV accumulation in tobacco leaves. (A), Leaf disks from healthy and infected leaves were pulse-labelled for 1.5 h with 3-H leucine to measure total protein synthesis (■), or with 3-H histidine to measure host protein synthesis (□). (B), Time course of TMV accumulation. From Fraser and Gerwitz, 1980.

4.2.1. Host protein metabolism in TMV-infected tobacco leaves

Fig. 4.1 shows changes in rates of protein synthesis during TMV accumulation in tobacco leaves. Total (host plus viral) protein synthesis was measured by incorporation of ^{3}H-leucine in pulse incubations, and showed no differences from uninfected control leaves for the duration of virus accumulation. The conclusion is that infection did not alter the overall host capacity for protein synthesis. To measure the rate of host protein synthesis alone, use was made of the fact that the coat protein of common isolates of TMV does not contain the amino acid histidine (Hennig and Wittman, 1972). Thus pulse labelling with ^{3}H-histidine labels only host proteins and virus-specified proteins other than coat. As the latter are synthesized in very small amounts (Sakai and Takebe, 1972), histidine incorporation essentially measures the rate of host protein synthesis. The results in Fig. 4.1 show that host protein synthesis was suppressed by as much as 75% during the period of most active virus multiplication, then rose to the control level when net TMV accumulation was over. This suggests that TMV coat protein synthesis occurred at the expense of host protein synthesis, but that there was no irreversible damage to the ability to produce host proteins.

The former conclusion contrasts with results from infected protoplasts: Siegel et al. (1978) reported that TMV-specific protein synthesis occurred in addition to, rather than at the expense of, host protein synthesis. However, protoplasts differ from whole plants in that the total amount of virus multiplication per cell tends to be about an order of magnitude lower than in whole leaves (Otsuki et al., 1972; Loebenstein et al., 1980) and must therefore represent much less competition.

Changes in total amounts of protein during virus multiplication showed a complex pattern, and depended heavily on the developmental stage of the leaf at the time of infection. Fig. 4.2 shows results for young and mature tobacco leaves. During development of the healthy leaf, total protein content increased during leaf expansion, then declined as the leaf aged. This was presumably integrated at the whole-plant level with a transport of protein turnover products from senescing lower leaves to growing upper leaves: compare the timings of the curves for healthy leaf proteins in Fig. 4.2 A and B.

Infection of mature leaves (Fig. 4.2 B) prevented the net loss of total protein during ageing; this was a result of accumulation of TMV coat protein, and overall loss of host protein was actually greater than in healthy leaves. This provides some evidence that TMV protein synthesis may also have occurred at the expense of faster turnover of host protein. Note also that the amount of TMV coat protein which was accumulated during virus multiplication was approximately equal to the net loss of host protein content, suggesting that host protein turnover was the only significant source of precursors.

The suggested diversion of the products of protein turnover to synthesis of virus coat protein, which remained in the lower leaves, most probably reduced the amount available for export to the younger

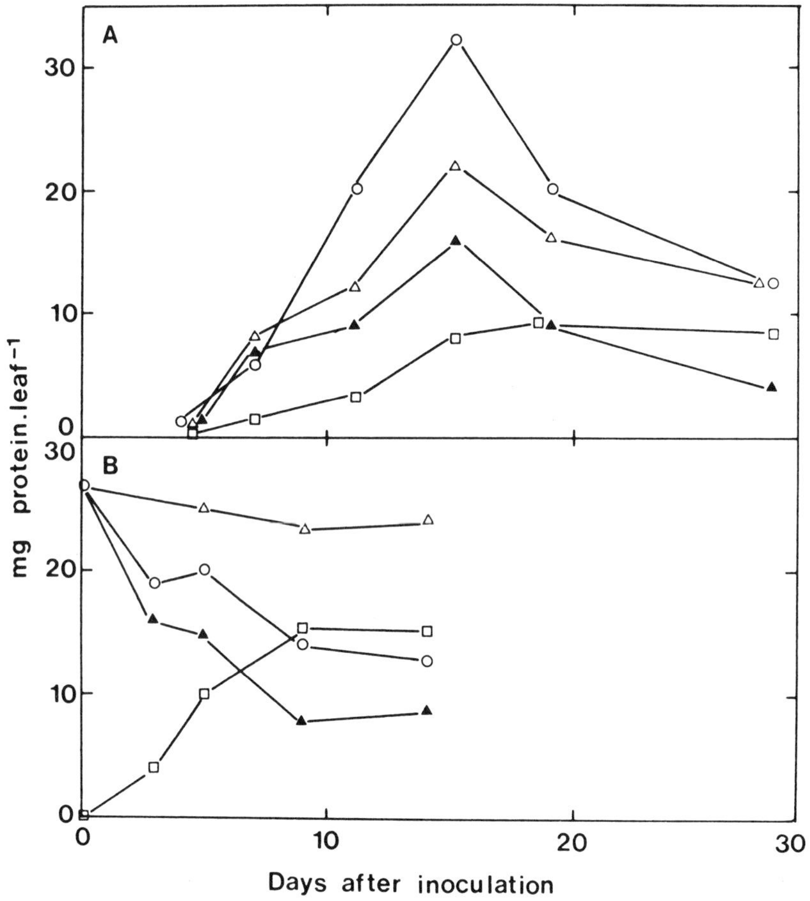

Fig. 4.2. Changes in host, viral and total proteins in tobacco leaves infected with TMV. (A), Leaves which became infected by systemic spread of virus when less than 5 mm long. (B), Leaves inoculated when they reached their final length of 230-250 mm. (□), TMV coat protein. (△), total protein in infected leaves. (▲), host protein in infected leaves. (○), total (host) protein in healthy leaves.

leaves. This is reflected in the lower overall (host plus virus) protein accumulation in young infected leaves than in comparative healthy leaves (Fig. 4.2 A). It is also clear that after allowing for accumulation of large amounts of coat protein, host protein accumulation was reduced to about 50% of that in healthy leaves. This could have been a significant cause of the observed poorer growth of young leaves on infected plants (see Chapter 7).

Thus, in sum, the effects of TMV infection on tobacco leaf protein metabolism involve the inhibition of host protein synthesis; a probable increase in the rate of host protein turnover, and a change in the pattern of distribution of protein through the plant with time, with possible consequences for host growth.

Sheen and Lowe (1979) and El-Meleigi et al. (1981) have compared the effects of TMV, TVMV and TEV on various crude protein fractions from tobacco cultivars with responses varying from hypersensitive lesion formation to systemic infection. Several differences emerged between different viruses and between cultivars, but most notable was a significant increase in soluble ('Fraction 2') protein stimulated by TVMV. This was partly at the expense of 'Fraction 1' (ribulose bisphosphate carboxylase oxygenase) and non-extractable proteins, but partly also due to net increase; it did not include viral coat protein.

There do not appear to have been similar types of 'global' quantitative studies for viruses which cause severe cytopathic effects, but do not accumulate to high concentrations, or for the viroids. This is perhaps surprising, since in these cases the interference from pathogen-coded protein metabolism is minimal or completely lacking. However, the effects of such pathogens on synthesis of particular proteins have been studied, especially for viroids, and these cases are considered below.

4.3. ELECTROPHORETIC ANALYSIS OF PROTEIN CHANGES AFTER INFECTION

Are all host proteins affected in the same way by infection, or do particular bands show differences? Are some proteins particularly strongly inhibited; are others synthesized at a faster rate? Can these changes be related to the development of pathogenic processes or host responses? These questions have been studied in a number of host-virus or host-viroid combinations.

4.3.1. Experimental approaches

A problem with analysis of total proteins from infected leaves is the vast number of individual proteins involved. One-dimensional electrophoresis may resolve up to 10^2 recognizable bands or shoulders, and two-dimensional techniques up to 10^3 or more (O'Farrell, 1975). This must be viewed against the cell complement of perhaps 10^4 to 10^5 individual polypeptides. Clearly, such techniques applied to total cell protein can only give an impression of changes in a small fraction of the number of protein species, and must reflect changes in those which are present in the largest amounts.

Paradoxically, the amount of information which advanced two-dimensional separations can provide is sufficient to require computer imaging and digitization techniques for comparisons. The amount of information which can be derived from electrophoretic techniques, and the breadth of the spectrum of proteins accessible to analysis, can also be expanded by use of additional fractionation methods such as pre-purification of sub-cellular organelles.

A technique commonly used to detect proteins newly synthesized after infection is to label healthy and infected tissues with amino acids or other protein precursors containing different radioactive isotopes, such as ^{3}H and ^{14}C. The proteins are extracted, mixed and fractionated together, and changes in the profile of proteins synthesized after infection are detected by increase or decrease in isotope ratio (e.g. Zaitlin and Hariharasubramanian, 1972; Jones and Reichmann, 1973; Singer and Condit, 1974; Ziemiecki and Wood, 1976). The limitation of the technique is that gels have to be sliced for dual channel scintillation counting. This technique has a lower resolving power for individual bands than autoradiography.

Another problem in electrophoretic analysis of proteins is to distinguish between changes in host components, and new proteins which are virus coded. Although the coding capacity of plant virus genomes is minute in comparison with the host, several factors increase the possibilities of confusing virus- and host-coded proteins. For several groups of plant viruses, the translation products _in vivo_ have not been fully identified, and translation of the viral RNA in vitro rarely provides a true spectrum of proteins (Davies and Verduin, 1979; Wilson and Glover, 1983). Furthermore, existence of 'readthrough' proteins and polyproteins which are cleaved to smaller products can make the total sequence length of all the virus-coded proteins in vivo considerably longer than predicted from the genomic length (reviewed by Davies and Hull, 1982).

Several techniques have been used to distinguish host- and virus-coded bands on gel fractionations. With infected protoplasts especially, ultraviolet irradiation has been used to suppress host protein synthesis (Paterson and Knight, 1975; Sakai and Takebe, 1974), and actinomycin-D to prevent transcription of DNA, and thus ultimately synthesis of host proteins (Jones and Reichmann, 1973; Ziemiecki and Wood, 1976). Chloramphenicol has been used to suppress host protein synthesis in chloroplasts (Paterson and Knight, 1975). Infection with different strains of the same virus, or with different pathogens, can also be useful for identification of host-coded changes (e.g. Tas and Peters, 1977; Camacho Henriquez and Sänger, 1982a). Viroids, which do not code for any known proteins, have been widely used to study changes in host proteins.

4.3.2. Observed changes

Despite the considerable number of investigations, and the diverse combinations of hosts and pathogens examined, these types of approach have given comparatively little information about host-virus interactions. In particular, most studies have described changes in a few proteins only, often with no idea of their function. We are still some way from thorough answers to questions such as the comparative numbers of host protein species which may be stimulated or depressed by infection, or what proportion of host proteins is unchanged. This information is an important pre-requisite to understanding possible controls of genomic and translational activity by virus infections.

In the case of viroid infections, Camacho Henriquez and Sänger (1982a) showed that PSTV increased the relative concentrations of nine host (tomato) proteins with apparent molecular masses from 10 kDa to 38 kDa, and decreased four proteins within the range 14.5 kDa

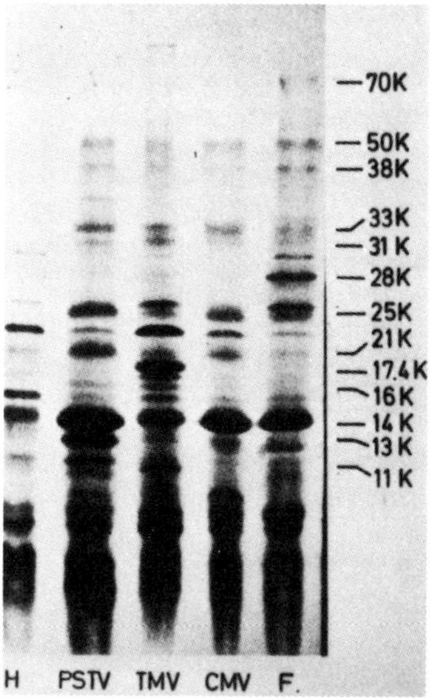

Fig. 4.3. Polyacrylamide gel separations of acid-extracted proteins from tomato plants: healthy (H), or infected with PSTV, TMV, CMV or Cladosporium fulvum (F). The apparent molecular weights of the proteins accumulating in infected tissue are shown on the right. Reproduced from Camacho Henriquez and Sänger, 1982b, by permission of the authors and Springer-Verlag.

to 105 kDa (Fig. 4.3). Similar changes were noted after TMV or CMV infection, suggesting that these formed a general physiological response to infection. The extent of the changes was related to the severity of visible symptoms. With PSTV infection of tobacco, Zaitlin and Hariharasubramanian (1972) reported synthesis of two very large proteins with molecular masses of 155 kDa and 195 kDa. Conejero and Semancik (1977) reported accumulation and enhanced synthesis of two small proteins with molecular masses of 13.7 kDa and 18 kDa in CEV-infected Gynura plants, and located them in the post-ribosomal supernatant fraction. Interestingly, two proteins of similar sizes appeared to increase when various other hosts were infected with this viroid (Conejero et al., 1979).

Singer and Condit (1974) studied the effects of infection by various mutants of TMV in tobacco. They found that the mutants suppressed synthesis of some host proteins, and caused new peaks, but no new protein was common to all TMV strains. This might be related to the diversity of visible symptoms caused by the different TMV mutants used.

CMV strains causing symptoms of different severity were used in studies of infection of cucumber (Ziemiecki and Wood, 1975; 1976) and tobacco (Roberts and Wood, 1981b). In cucumber, mild and severe strains caused similar changes, whereas in tobacco, it was suggested that the severe symptom strain caused decreases in several host proteins which were unaffected by the mild strain, although both strains multiplied to similar extents.

Tas and Peters (1977) reported that both CMV and TSWV stimulated accumulation of a 22 kDa host glycoprotein in cucumber cotyledons.

In wheat and barley infected with WSMV or BSMV and examined by 2-dimensional electrophoresis, White and Brakke (1983) found that some proteins were decreased, and others increased, as a result of infection. Those proteins which were increased were different for the two viruses, and most probably included coat and other virus-coded polypeptides. Many of the proteins which were decreased appeared to be similar for the two viruses.

Further examples of host protein changes induced by virus infection are considered in the narrower context of the stress or 'pathogenesis-related' proteins later in this Chapter.

4.4. CHANGES IN ENZYMES OR PROTEINS WITH KNOWN FUNCTIONS

Some aspects of changes in enzyme activities are considered in more detail in the Chapters on respiration, photosynthesis and resistance. In this Chapter, we are concerned with effects of infection on particular enzymes or structural proteins primarily as a means of examining the interactions between virus multiplication and host protein metabolism.

4.4.1. Proteins associated with the chloroplast

There are several reasons for studying the effects of virus infection on the metabolism and activity of ribulose bisphosphate carboxylase-oxygenase (Rubisco; Fraction 1): it has a central role in photosynthetic CO_2 fixation and in photorespiration; it is the commonest protein in plants and thus easy to study; its structure and genetics pose particularly interesting problems. The small subunit is coded in the nuclear genome, synthesized on cytoplasmic ribosomes, and transported into the chloroplast with the assistance of a transit polypeptide. The large subunit is coded and synthesized within the chloroplast (Ellis, 1981b). In view of the reported differential effects of virus infections on chloroplast and cytoplasmic ribosomes, and on chloroplast protein synthesis (Hirai and Wildman, 1969; Fraser, 1969), it is perhaps surprising that the questions which virus infection may pose about the control of Rubisco synthesis and transport do not seem to have been examined directly.

The few studies of Rubisco synthesis appear to show that it is inhibited early in infection, for example of tobacco by TMV (Doke and Hirai, 1970; Mohammed and Randles, 1972), and of Chinese cabbage by TYMV (Crosbie and Matthews, 1974a). Mohammed and Randles (1972), however, showed that inhibition of Rubisco synthesis in TSWV-infected tobacco did not occur during the period of mosaic symptom development and most active virus accumulation, but was delayed until the later phase of veinal necrosis, when virus content was in decline.

The effects of infection on the carboxylase and oxygenase activities of Rubisco, and on possible compensatory metabolism, are considered in Chapter 5.

Detailed studies of the effects of infection on other chloroplast proteins have now started to appear. Naidu et al. (1984a) studied the chlorophyll-protein complexes (Anderson, 1982) of PGMV-infected peanut leaves, in which a severe mosaic phase is followed by a recovery phase. In the severe phase, infected leaves had a higher content of the CPI complex of photosystem I, whereas other light-harvesting complexes, and especially the photosystem II reaction centre complex, were reduced. Thylakoid membranes from leaves with severe mosaic showed increases in some polypeptide bands, and decreases in others. The polypeptide and protein-chlorophyll patterns of infected leaves in the recovery phase resembled those of healthy leaves.

Platt *et al.* (1979) also showed a lower content of chlorophyll (*a*+*b*) light-harvesting protein in variegated leaves of *Tolmiea menziesii* with a double infection by TBSV and CMV, although no recovery phenomenon was reported.

4.4.2. Enzymes of peroxide metabolism

Catalase and peroxidase are both capable of metabolizing hydrogen peroxide, which is potentially cytotoxic. However, their role in metabolism of healthy tissue has long been in doubt, and while changes in enzyme activities in virus-infected tissues have been recorded, the possible role of hydrogen peroxide in necrogenesis especially has received little direct attention. Peroxidase, as well as polyphenoloxidase, also catalyses the oxidation of phenolic compounds to quinones, which then polymerize to form the brown pigments associated with lesions, and which may be implicated in necrogenesis (Van Loon, 1983a).

Recently, evidence has been obtained that a hypersensitive reaction of potato protoplasts to infection by incompatible races of *Phytophthora infestans* is associated with increased generation of the highly toxic superoxide (O_2^-) anion, which is converted to hydrogen peroxide by superoxide dismutase (Doke, 1983a; 1983b). This type of metabolic change does not seem to have been investigated in viral infections, but as some of the membrane changes reported for the potato work were similar to those occurring in the hypersensitive reaction to virus infections (reviewed in Chapter 6), the question might be worthy of examination.

Increases in peroxidase activity have been reported for many host/virus combinations. Generally, infections leading to a hypersensitive response seem to cause greater increases than those giving a systemic mosaic (Sheen and Diachun, 1978; Wagih and Coutts, 1982a; Van Loon, 1982). Although stimulated peroxidase activity was detectable in uninfected parts of tobacco plants (Simons and Ross, 1971a; Van Loon and Geelen, 1971), the greatest increases occurred in cells around the local lesions caused by TMV (Weststeijn, 1976). These increases in enzyme activity appeared to be a general response to necrotic damage, as peroxidase activity was also stimulated by heat-induced lesions (Balazs *et al.*, 1977).

There is disagreement on whether increased activity results from synthesis of new peroxidase isoenzymes, or from more of existing types. Systemic infection of cucumber with CMV (Wood, 1971) stimulated accumulation of isoenzymes found in healthy plants, and the degree of the increase was related to symptom severity and amount of virus multiplication (Wood and Barbara, 1971). Novacky and Hampton (1968) also excluded induction of new isoenzymes for a number of host-virus combinations. In contrast, Solymosy *et al.* (1967) reported new isoenzymes in *Phaseolus vulgaris* and *Nicotiana glutinosa* inoculated with various viruses, and Farkas and Stahmann (1966) found that SBMV infection of bean caused premature appearance of a senescence-related peroxidase. De Zoeten and Rettig (1972) used isoelectric focussing to show a new peroxidase band in PEMV-diseased peas. The higher resolution of this technique, compared with electrophoresis on non-denaturing gels used in most other studies, probably helped to resolve the new band.

Catalase activity has been reported to be reduced slightly by systemic infection, for example of pumpkin by WMV (Singh, 1983), but to be increased by necrotic infection of tobacco by TMV (Simons and Ross, 1971a).

Matkovics *et al.* (1978) reported that TMV infection reduced superoxide dismutase activity in tomato. Curiously, this occurred in plants containing the *Tm-1* or *Tm-2*^{2} resistance genes. The latter allows almost no virus multiplication (Pelham, 1972).

4.4.3. Polyphenoloxidase

The quinones formed by polyphenoloxidase activity have been shown in some cases to have antiviral activity (Saksena and Mink, 1970) (see Chapter 6). Several studies have shown increased polyphenoloxidase activity after the appearance of necrotic lesions, for example with TMV-infected tobacco (Van Kammen and Brouwer, 1964; John and Weintraub, 1967; Simons and Ross, 1971a); CCMV-infected soybeans (Batra and Kuhn, 1975) and TNV-infected cowpea and cucumber (Wagih and Coutts, 1982b). Generally, increased activity was localized near the site of infection, although increases in non-infected leaves have been reported (Van Kammen and Brouwer, 1964; Batra and Kuhn, 1975). In contrast, the necrotic reactions of beans to AMV, and of *Chenopodium amaranticolor* to TNV, did not involve increases in polyphenoloxidase (Vegetti *et al.*, 1975; Faccioli, 1979). A small increase in polyphenoloxidase activity has been found in systemically-infected tobacco (Van Loon, 1982).

4.4.4. Nucleases and proteases

Wyen *et al*. (1972) found that the increased level of RNase activity in tobacco hypersensitive to TMV was associated with a purine-specific form of the enzyme. Activity of other nucleases was little altered by infection or mechanical injury, and there was no correlation between virus multiplication and nuclease level.

Changes in RNase isoenzyme patterns and activities were studied by Wagih and Coutts (1982a, 1982b) in cowpea leaves and cucumber cotyledons infected with TNV. Infection caused up to twice as much RNase activity in both hosts, but there was no clear evidence of synthesis of new or particular isoenzymes. Osmotic stress also stimulated RNase activity, leading the authors to suggest that the increase was a general stress response and not specifically stimulated by necrotic infection.

Day (1984) found no differences in total RNase or protease activities after systemic infection of bean leaves by BCMV.

4.4.5. Enzymes of the phenylpropanoid pathway

Phenylalanine is the immediate precursor of the phenylpropanoid pathway, which serves among other things as a source of precursors for lignin synthesis. Ultrastructural evidence suggests that the hypersensitive resistance reaction to virus infection involves lignification, and this has stimulated interest in activation of the phenylpropanoid pathway after infection. Evidence that lignification is involved in resistance is considered in Chapter 6.

Increased activity of the first enzyme of the pathway, phenylalanine-ammonia lyase (PAL), was detectable before the appearance of necrotic symptoms of TMV on tobacco (Simons and Ross, 1971b; Fritig et al., 1973). Later experiments using density labelling showed that the increased activity represented newly synthesized enzyme, and not activation of existing enzymes (Duchesne et al., 1977).

Subsequent enzymes in the pathway, cinnamic acid-4-hydrolase (CAH) and caffeic acid-O-methyl transferase (OMT), became activated in sequence (Legrande et al., 1978), and density labelling experiments again showed that this was due to new synthesis (Collendavelloo et al., 1983). These meticulous experiments are among the most detailed studies of changes in enzyme metabolism following virus infection.

4.5. PATHOGENESIS-RELATED PROTEINS

Early studies of protein composition in virus-infected plants used gel electrophoresis under non-denaturing conditions to examine changes accompanying the development of TMV lesions on hypersensitively-reacting tobacco. These experiments quickly showed major differences; up to four new host-coded proteins were found as a result of infection (Gianinazzi et al., 1970; Van Loon and Van Kammen, 1970). These proteins, named 'b-proteins' and later 'pathogenesis-related' (PR) proteins, were unusual in having a particularly high electrophoretic mobility under the conditions used. They were therefore comparatively easy to study, in that they could readily be separated from the bulk of the other cell proteins. Subsequently, the discovery that PR proteins are soluble at pH 2.8 (Van Loon, 1976; Gianinazzi et al., 1977) allowed a simple enrichment step during extraction.

A second source of stimulus to the study of PR proteins came from the suggestion that they might be involved in the acquired systemic resistance of plants to a second infection, perhaps in a manner similar to interferon in mammalian cells (Kassanis et al., 1974; Gianinazzi, 1982). This possible involvement in resistance is considered in more detail in Chapter 6.

PR-like proteins have now been identified in a number of plant species, and their induction and properties have probably been studied more extensively than any other host protein in the context of virus infection. Despite this, their function remains enigmatic.

4.5.1. Occurrence of PR proteins

When is a protein a PR protein? The common feature is absence from normal, healthy plants, and induction by viral and other pathogens. But within and especially beyond this, the definition becomes somewhat blurred. Thus under certain circumstances, PR proteins can be detected in untreated, healthy plants, and they can be induced by many factors other than pathogens. Where data are available, PRs from different sources have tended to share some common features: acid solubility, high electrophoretic mobility on non-denaturing gels and protease resistance. However, these are by no means universal features, and some are perhaps more illusory than real.

The 'type species' for PR proteins are those found in tobacco, and the most extensive studies of their occurrence in the genus Nicotiana have been conducted by Gianinazzi and co-workers. Ahl et al. (1982) and Gianinazzi and Ahl (1983) found evidence for at least seven different PRs in various species, with evidence for intra- and inter-specific variation. Further studies using trypsin treatment to remove other host proteins have revealed up to seven further PR proteins, although these have not yet been fully characterized (Fig. 4.4).

PR-like proteins have also been detected in cowpea and cucumber (Coutts and Wagih, 1983); tomato (Camacho Henriquez and Sänger, 1982b); Gomphrena globosa (Pennazio and Redolfi, 1980); Phaseolus bean (Abu-Jawdah, 1982) and several other species (reviewed by Redolfi, 1983 and Van Loon, 1985).

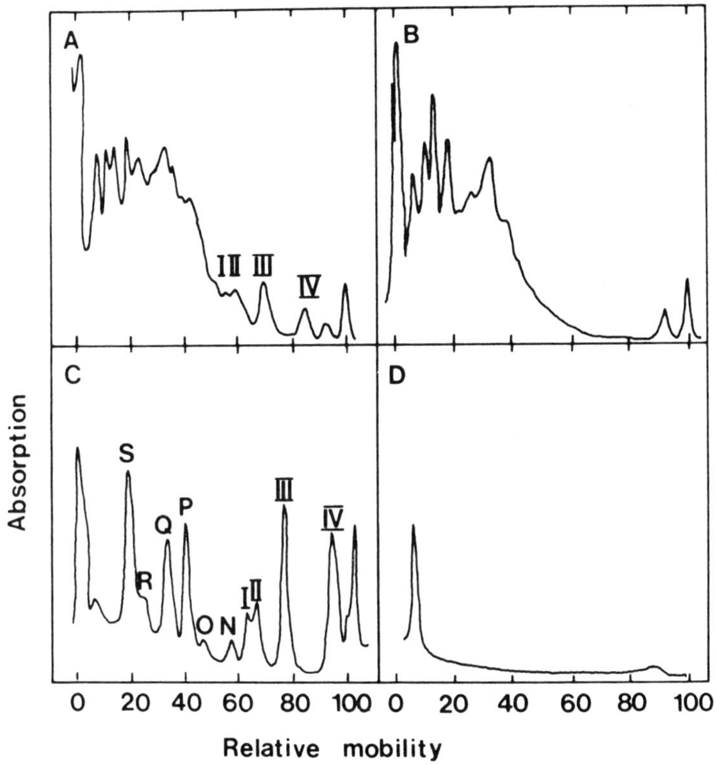

Fig. 4.4. Densitometer tracings of electrophoretic patterns in 10%, non-denaturing polyacrylamide gels of proteins from tobacco cv. Samsun NN, 7 days after inoculation with TMV (A,C) or water (B,D). (A and B), subjected to electrophoresis immediately after extraction. The PR proteins in the infected tissue are indicated I-IV. (C and D), incubated for 24 h with proteinase K before fractionation. Additional PR proteins revealed in (C) are indicated by letters. Reproduced from Van Loon (1983a) by permission of the author and Academic Press.

4.5.2. Properties of PR proteins

The vast majority of PR proteins which have been characterized to date have molecular masses within the range 10 to 20 kDa. The high electrophoretic mobility of the first PR proteins studied was a consequence of a high proportion of acidic amino acid residues (Antoniw and White, 1983). However, trypsin treatment has now

74

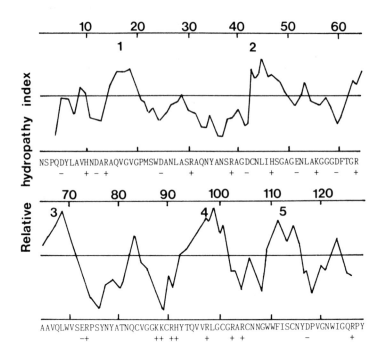

Fig. 4.5. Amino acid sequence of the tomato pathogenesis-related protein p14, shown in the single letter amino acid code (IUPAC–IUB Commission, 1968). Also plotted is the relative hydropathy index, to show the five hydrophobic domains of the molecule. Reproduced from Lucas et al. (1985) by permission of the authors and IRL Press, Oxford.

revealed PR proteins in tobacco with much lower electrophoretic mobilities (Van Loon, 1982; Pierpoint et al., 1981), which may well turn out to have a much lower proportion of negatively charged residues. The tomato protein designated as p14 is highly basic (Camacho Henriquez and Sänger, 1984).

Amino acid compositions have been determined for the tomato p14 protein, and for five tobacco PR proteins, and all contain high proportions of apolar residues (Camacho Henriquez et al., 1983; Antoniw and White, 1983). This property might suggest an association with membranes (Benson and Jekela, 1976). The p14 protein has been sequenced (Lucas et al., 1985), and shows five hydrophobic domains

(Fig. 4.5), leading Lucas et al. (1985) to suggest association with membranes.

The PR1a protein has been partially sequenced by the same workers. They showed sequence similarities between p14 and PR1a. Furthermore, sequence comparisons with over 3,000 other proteins showed no similarities, suggesting that PR proteins are of a unique structural type.

Three of the tobacco PR proteins, designated PR1a, b and c, have been shown to have very similar molecular weights and amino acid compositions; it is suggested that they are charge isomers of a single protein (Antoniw et al., 1980; Antoniw and White, 1983).

The early tobacco PR proteins were shown to contain no carbohydrate (Van Loon, 1982). However, PR proteins from some other sources have been found to give a positive reaction in Schiff staining (Tas and Peters, 1977; De Wit and Bakker, 1980; Andebrhan et al., 1980) and are thus likely to be glycoproteins. Pierpoint (1983a) showed that some of the more recently discovered tobacco PR proteins would bind to chitin, and stained in the Schiff reaction. This led him to suggest that these PR proteins might be acetylglucosamine-specific lectins.

Protease resistance has also been reported for bean PR proteins (Redolfi, 1983), but Camacho Henriquez and Sänger (1984) found that tomato p14 was protease sensitive. Protease sensitivity of PR proteins from numerous other sources does not seem to have been tested.

Apart from the possible functions mentioned above for particular PR proteins, and their role in resistance which is questioned in Chapter 6, various other functions have been examined. Van Loon (1982) and Antoniw and White (1983) recorded 25 enzyme activities which were not associated with the four best-characterized tobacco PR proteins. Redolfi (1983) has compared known PR proteins with protease inhibitors, heat- and osmotic-shock proteins, and other proteins with a suggested role in resistance, without finding any strong parallels. One interesting finding which may give a clue to the function of these proteins is that they seem to occur almost exclusively in the cell walls or intercellular spaces (Parent and Asselin, 1984) and are very difficult to detect in protoplasts prepared from leaves containing the proteins (Wagih and Coutts, 1981). Occurrence of PR proteins in tobacco leaves also seems to be associated with increased formation of multilammellar structures between the cytoplasm and cell wall (Fraser and Clay, 1983). Although these would suggest wall and membrane changes, their association with PR proteins remains questionable, and multilammellar structures are also found in healthy plants (Appiano and D'Agostino, 1985).

4.5.3. Induction of PR proteins

PR proteins are normally absent from, or difficult to detect in, normally growing, healthy plants (Van Loon, 1976; Redolfi, 1983), although several exceptions to this have now been reported (e.g. Tas and Peters, 1977; Fraser, 1981; Wagih and Coutts, 1982a). Large amounts of proteins are induced by viruses which cause necrotic infections, but systemic infection can also be an effective inducer (Kassanis et al., 1974). Necrosis is thus not required for induction, but the highest production of PR proteins does seem to occur in necrotic infections, and the highest accumulation of PR protein within the plant occurs in tissue near the lesions (Van Loon, 1983b; Rohloff and Lerch, 1977).

Representative examples of PR-inducers are presented in Table 4.1; more comprehensive listings of viral and other inducers can be found in the reviews by Van Loon (1983b) and Fraser (1985). It is clear from Table 4.1 that fungal and bacterial pathogens, and several viroids, induce PR proteins. Furthermore, a wide variety of chemicals and developmental factors are also effective inducers. These results strongly suggest that PR proteins are a general response to stress, and are not specifically and intimately related to any particular viral pathogen.

In tobacco, Fraser (1981) showed that PR proteins accumulated in all leaves of healthy plants as they commenced monocarpic senescence. Removal of either the developing inflorescence, or the senescing lower leaves, inhibited the accumulation of PR proteins in the

Table 4.1. Inducers of pathogenesis-related proteins.

Host	Inducer	References
	VIRUSES	
Tobacco cv. Samsun-NN	TMV	Van Loon and Van Kammen, 1970
Tobacco cv. Xanthi-nc	TMV	Gianinazzi *et al.*, 1970
Tobacco cv. Xanthi-nc	PVY,CMV,PVX,AMV	Kassanis *et al.*, 1974
Tobacco cv. Samsun-nn	TNV	Van Loon and Dijkstra, 1976
Cucumber, cowpea	TNV	Coutts and Wagih, 1983
Tomato	TMV,CMV	Camacho Henriquez and Sänger, 1982b
Bean	AMV,PSV	Abu-Jawdah, 1982
	VIROIDS	
Gynura aurantiaca	CEV	Conejero and Semancik, 1977
Tomato	PSTV	Camacho Henriquez and Sänger, 1982b
	MICROORGANISMS	
Tobacco	*Pseudomonas syringae*	Ahl *et al.*, 1981
Tobacco cv. Burley	*Thielaviopsis basicola*	Gianinazzi *et al.*, 1980
	DEVELOPMENTAL FACTORS	
Tobacco cv. Xanthi-nc	callus culture	Antoniw *et al.*, 1981
Tobacco cv. Xanthi-nc	flowering	Fraser, 1981
N. glutinosa	nutrient deficiency	Ahl and Gianinazzi, 1982
	CHEMICALS	
Tobacco cv. Xanthi-nc	polyacrylic acid	Gianinazzi and Kassanis, 1974
Tobacco cv. Xanthi-nc	aspirin	White, 1979
Tobacco cv. Samsun-NN	salicylic, benzoic acids	Van Loon and Antoniw, 1982
Tobacco cv. Samsun-NN	ethephon	Van Loon, 1977
Tobacco cv. Xanthi-nc	auxins, cytokinin	Antoniw *et al.*, 1981
Cucumber	mannitol	Wagih and Coutts, 1981
Tobacco cv. Xanthi-nc	carbendazim	Fraser, 1982

remaining leaves by about 90%. This suggests that induction depends on signals emanating from both ends of the shoot; a possible model is that a primary signal from one end induces a secondary signal from the other, which is the mobile inducer of PR protein synthesis.

Gianinazzi and Ahl (1983) provided evidence for a mobile inducer using a totally different experimental approach. They grafted non-infected scions from n-gene (non-necrotic) tobacco cultivars on to rootstocks of N-gene (necrotic) cultivars with a localized TMV infection. The stock and scion were chosen to have different PR protein patterns. PR proteins, with the pattern characteristic of the stock genotype, were induced in the stock, although no virus was transmitted across the graft union. This experiment also suggests that the mobile inducer is not specific for PR protein genes.

Ahl and Gianinazzi (1982) further showed that the interspecific hybrid N. glutinosa (NN) x N. debneyi (nn) produced PR proteins constitutively without inducers, and that the rootstock of the hybrid would induce PR proteins characteristic of the scion in grafting experiments. It is assumed that the mobile inducer is produced constitutively in the hybrid. It may be relevant that the hybrids were smaller than the parents and appeared to show early senescence-like symptoms, suggesting that the constitutive inducer might resemble the mobile inducer occurring in monocarpic senescence (Fraser, 1981).

The biochemistry of induction of PR proteins by natural and applied chemicals has been extensively studied by Van Loon, who has suggested a central role for ethylene (Van Loon, 1983b). Fig. 4.6 is a diagrammatic representation of some of the factors and inhibitory influences which may control PR protein synthesis. Ethephon, an ethylene-releasing compound, and 1-amino cyclopropane-1-carboxylic acid (ACC), the natural precursor of ethylene, are both effective inducers of PR proteins in tobacco. Almost all other biotic or synthetic inducers of PR proteins also induce or stimulate ethylene biosynthesis. Furthermore, incubation of TMV-infected tobacco leaf disks with aminoethoxyvinylglycine, an inhibitor acting on the ethylene biosynthetic pathway, blocked ACC formation and accumulation

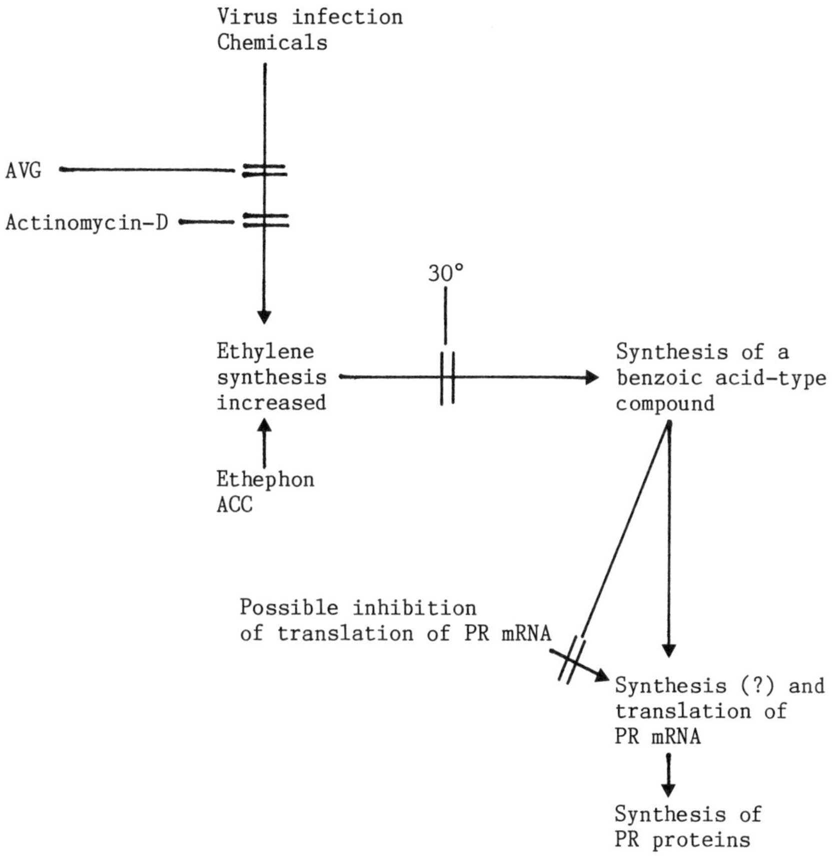

Fig. 4.6. A possible control pathway for synthesis of PR proteins. == indicates an inhibition. AVG, aminoethoxyvinylglycine; ACC, 1-amino cyclopropane-1-carboxylic acid. Actinomycin-D or the putative benzoic acid-type compound remove the inhibition of PR mRNA translation. Based on Van Loon and Van Kammen (1970); Carr _et al_. (1982); Van Loon and Antoniw (1982) and Van Loon (1983b).

of PR proteins.

However, one group of compounds induced PR proteins without stimulating ethylene biosynthesis; this included benzoic acid and some of its derivatives such as 2-hydroxybenzoic acid (salicylic acid) and 2,6-hydroxybenzoic acid (Van Loon and Antoniw, 1982). It appears that PR protein induction requires a hydroxyl group exclusively at the ortho position.

A further distinction between ethylene- and benzoic acid-induction of PR proteins was that the former did not occur at temperatures above 30°C, whereas salicylic acid would induce PR proteins at 32°C (Van Loon and Antoniw, 1982). These findings led Van Loon (1983b) to postulate that ethylene induces a benzoic acid-type compound in the plant by a temperature-sensitive step, and that this compound is then the direct inducer of PR protein synthesis by a temperature-insensitive step. The suggested natural, benzoic acid-type inducer has not yet been identified in tobacco, although similar compounds are known to occur in other plants. It will be interesting to examine whether the proposed inducer bears any resemblance to the proposed mobile inducer in senescent or grafted plants.

The complexities of the control of PR protein biosynthesis were increased with reports by Carr _et al_. (1982) and Carr (1985) that a translational level might be involved. _In vitro_ translation of RNA from healthy, untreated tobacco leaves showed synthesis of PR proteins which were not detectable in leaf extracts. Matsuoka and Ohashi (1986) provided evidence supporting control of synthesis of PR proteins at translation. But probing healthy leaf total RNA with cDNAs for PR1 proteins failed to detect significant levels of PR1 mRNA (Hooft van Huijsduijnen _et al_., 1985), and the amount of PR1 mRNA in healthy leaves therefore remains in some doubt. Perhaps it is highly dependent on plant age and environmental conditions.

Carr (1985) also reported that actinomycin-D, an inhibitor of transcription, stimulated PR protein accumulation in healthy leaves. He suggested that an unstable transcription product might inhibit translation of stable PR mRNA. After addition of actinomycin-D, the transcription product would not be synthesized, its concentration in the cytoplasm would decline, and the inhibition of translation of stable PR mRNA would be relaxed.

In contrast to this work, earlier reports suggested that actinomycin-D inhibited accumulation of PR proteins in TMV-infected leaves (Van Loon and Van Kammen, 1970), or during chemical induction

(Kassanis *et al.*, 1974). However, these inhibitions were small, and actinomycin-D did not inhibit induction of PR proteins by ethylene or salicylic acid (Van Loon and Antoniw, 1982; Van Loon, 1983b). Actinomycin-D did, however, inhibit the production of ethylene in TMV-infected tobacco (De Laat and Van Loon, 1983), by an amount similar to the inhibition of PR protein accumulation. These results led Van Loon (1983b) to suggest that the transcription-dependent (actinomycin-D sensitive) step in the control of the induction of PR protein must occur before the stimulation of ethylene biosynthesis (Fig. 4.6).

Finally, Hooft van Huijsduijnen *et al.* (1985) showed by sequencing of the 5'-terminal regions of PR1 mRNAs that the PR1 proteins are derived from a precursor by removal of an N-terminal polypeptide 30 amino acids long. They suggested that this might act as a signal polypeptide, perhaps for transport.

4.6. AMINO ACID AND NITROGEN METABOLISM

4.6.1. Amino acids and amides

A number of early studies followed changes in concentrations of amino acids and amides after infection. Commoner and Nehari (1953) showed reduced concentrations of these components in leaf disks floated on nutrient solution, during the period of active TMV multiplication. However, Porter (1959) argued that these conclusions could not be extrapolated to intact plants. The majority of studies have shown increases in most amino acids and amides, although a few compounds decreased after infection (Selman *et al.*, 1961; Singh and Mohan, 1982; Randles and Coleman, 1972). Hayashi (1962) found that tobacco leaves infected with TMV had a 50% higher amino acid-activating activity than controls.

Harpaz and Applebaum (1961) listed 11 virus diseases which caused an accumulation of asparagine, and suggested that this might play a central role in symptom formation. Bozarth and Diener (1963) studied a range of amino acids and amides in tobacco plants infected singly or doubly with PVX and PVY, thus causing variable symptoms, but were unable to relate changes directly to symptom formation.

Mohanty and Sridhar (1982) found that RTV-infection of rice caused higher proline concentrations. Proline also increased in detached, senescing leaves, or in leaves treated with abscisic acid. The last observation is of interest in view of the possible links between virus infection and abscisic acid metabolism discussed in Chapter 7.

It is probable that amino acid and amide concentrations are increased because of reduced synthesis or increased turnover of host proteins. In most infections the amount of virus protein synthesized may not be sufficient to absorb the amounts of precursors available, except for viruses such as TMV which multiply to very high levels. Indirect evidence bearing on the relationship between TMV and host protein metabolism has been considered earlier in this Chapter.

4.6.2. Nitrogen fixation

There have been several studies of the effects of virus infection on nitrogen fixation by leguminous plants. DEMV was reported to increase nodulation and nitrogen fixation of field bean (Dolichos lablab) (Rajagopalan and Raju, 1972). However, the majority of other studies have found inhibitory effects on various stages of nodulation and fixation. Thus Khadhair et al. (1984) found up to 37% fewer nodules, 60% less nitrogenase activity, and 60% less leghemoglobin, in WCMV-infected red clover. Reductions of similar orders of magnitude have been found in soybeans infected with SMV (Tu et al., 1970a; 1970b) or TRSV (Orellana et al., 1980; 1982), and peanut infected with PMV (Wongkaew and Peterson, 1983). Most reports showed that infectious virus was recoverable from the impaired nodules, that the greatest impairment of nodulation and fixation occurred in plants inoculated at an early stage of growth, and that viral inhibition of some factors, such as leghemoglobin accumulation, diminished or disappeared as the plant aged. This recovery was, however, too late to contribute to the nitrogen economy of the nodule which was senescent by then.

The evidence suggests that viruses might impair nodulation and nitrogen fixation at several stages. Tu *et al.* (1970a) found that healthy soybeans were always more susceptible to *Rhizobium* infection than SMV-infected plants. Khadhair *et al.* (1984) recovered significantly fewer viable bacterial colonies from WCMV-infected nodules, and suggested that the virus may have inhibited rhizobial multiplication, or transformation into bacteroids. They also suggested that the suppression of nodulation, and the generally smaller nodules on infected plants, may have resulted from interference with auxin production in the developing nodule. They ascribed the reduced nitrogen fixing ability of infected nodules to two indirect effects: possible reduction in energy supply as a consequence of viral inhibition of photosynthesis, and oxygen imbalance caused by inhibition of leghemoglobin accumulation.

4.6.3. Nitrate reductase

Nitrate reductase activity has been found to be increased in WMV-infected pumpkin (Singh, 1983) and CPMV-infected cowpea (Singh and Singh, 1982). Khadhair *et al.* (1984) found increased nitrate reductase activity in root nodules of WCMV-infected red clover. They suggested further that since nitrite, the product of nitrate reductase, can inhibit nodular nitrogenase activity, the virus-stimulated utilization of nitrate might be an important inhibitor of nodular nitrogen fixation.

4.7. CONCLUSION

The effects of virus infection on host protein metabolism have offered opportunities for diverse investigations. Problems remain in deciding which processes are directly involved in pathogenesis, and which are merely secondary consequences. Modern molecular techniques are permitting more precise investigation of the structure and metabolism of particular proteins associated with infection, although studies of function continue to lag behind. However, the ability to use nucleic acid sequences to predict amino acid sequences, and to use these for the construction of synthetic polypeptides, offers a new and highly specific means of preparing antibodies to particular proteins. These methods should eventually open up new approaches to the study of function.

CHAPTER 5

Respiration and Photosynthesis in the Virus-Infected Plant

5.1. RESPIRATION

5.1.1. Respiration in leaves with a hypersensitive response

Almost all investigations have been with Nicotiana species forming lesions after inoculation with TMV. In N. glutinosa, Yamaguchi and Hirai (1959) found an increase of up to 1.4 times in oxygen uptake, commencing when lesions appeared. They suggested that the increased respiration was related to the necrosis, rather than to virus multiplication which was detectable earlier. Sunderland and Merrett (1965), however, found that the increase in respiration started before lesion appearance in N. tabacum cv. Xanthi-nc.

In N. glutinosa, Weintraub et al. (1972) found that the numbers of mitochondria rose at the time of lesion appearance to 50% more than in healthy plants, but within ten hours after this had fallen to only half the frequency in the controls. They associated the transitory increase with observed increases in mitochondrial enzyme activities (Weintraub et al., 1964), and the subsequent decline with general tissue disruption during necrogenesis.

Sunderland and Merrett (1964) found small increases in the concentrations of ATP and ADP at a late stage of lesion development, while Solymosy and Farkas (1962; 1963) reported three-fold increased activity of glucose-6-phosphate dehydrogenase and greater gluconate-6-phosphate dehydrogenase activity. Pentoses accumulated; the authors suggested that this was due to activation of the early stages of the hexose monophosphate shunt, with the effect that the later stages of the pathway could not cope with the increased flux.

It is clear from all these reports that the necrotic response is accompanied by a general activation of respiratory activity; this appears to be located in the tissue surrounding the lesions.

5.1.2. Respiration in systemically-infected tissues

A wider range of hosts and viruses has been examined than with necrotic infections. In some, infection had little or no effect on respiration. These include BYV-infected sugar beet (Hall and Loomis, 1972a) and BYDV-infected wheat (Jensen, 1972). In BYDV-infected barley, respiration was not affected when measured per unit dry weight, but was higher on a per unit fresh weight basis, because infection increased the proportion of dry matter (Jensen, 1968).

In other host-virus combinations, infection has been reported to increase respiration. MDMV infection of corn caused a 32% increase, associated with the appearance of mosaic symptoms (Tu et al., 1968). Respiration increased on infection of CMV-susceptible cucumbers, but not in a CMV-resistant variety (Mencke and Walker, 1963). The latter was symptomless, but accumulated large amounts of virus, suggesting that increased respiration was associated with development of mosaic symptoms, rather than with virus multiplication as such.

In tobacco, activities of certain enzymes of glycolysis and of the pentose phosphate cycle were increased during the period of most active TMV accumulation (Makovcova and Sindelar, 1977). Leal and Lastra (1984) found respiration 80-100% above healthy plant levels in ToYMV-infected tomato.

5.2. PHOTOSYNTHESIS

Studies of the effects of virus infection on photosynthesis have been more comprehensive and wide-ranging than those on respiration. They have included research on overall rates of photosynthesis and photorespiration, studies of the photobiology of energy capture and processing, and effects on chloroplast ultrastructure and chlorophyll metabolism. Finally, longer-term and whole-plant experiments have considered virus effects on photosynthesis and respiration, in terms of the gross carbon economy of the plant. The effects of infection on growth and yield will be discussed in Chapter 7. The effects of virus infection on synthesis of certain proteins involved in photosynthesis were considered in Chapter 4.

5.2.1. Rates of photosynthesis and photorespiration

In almost all hosts studied, virus infection reduced the net photosynthetic rate. Table 5.1 summarizes a number of examples, which include plants with both the C_3 and C_4 pathways of CO_2 fixation (reviewed in Hatch, 1976).

When comparing different investigations, care has to be taken that photosynthetic rate has been expressed on a comparable basis. Thus most workers have measured photosynthesis per unit leaf area, but some have expressed the results per unit fresh or dry weight. Expression per unit chlorophyll is perhaps especially subject to confusion, as there is evidence that viral infection may alter the distribution and activity of chlorophyll, as well as causing changes in chlorophyll concentration (discussed in section 5.2.2).

Most investigators have measured differences in net photosynthetic rate at saturating light intensities, and some have found smaller inhibitory effects of infection at sub-saturating light intensities. The light history of the experimental plants may also condition the response: Hall and Loomis (1972a) found greater inhibition of net photosynthesis at saturating light intensity in BYV-infected beet plants grown under high light intensities, than in those grown under moderate illumination.

Fig. 5.1 shows the effects of PGMV on photosynthesis in peanut leaves at three stages of symptom development (Naidu et al., 1984b). During the early stage with mild chlorotic spots, net photosynthesis was reduced at all light intensities. The reduction due to infection became greatest during the middle, severe symptom stage. In the third stage, where leaves had recovered from severe symptoms, net photosynthesis returned to the healthy plant level, indicating a

Table 5.1. Inhibition of photosynthesis by viral infections.

Host	Virus	Percentage reduction in photosynthesis	References
Maize	MDMV	25	Tu _et al._, 1968
Barley	BYDV	20	Jensen, 1968
Wheat	BYDV	45	Jensen, 1972
Tomato	ToYMV	50	Leal and Lastra, 1984
Tomato	TAV	40	Hunter and Peat, 1973
Sugar beet	BYV	30	Hall and Loomis, 1972a; 1972b
Peach	PNRV/PDV	13	Smith and Neales, 1977a
Peanut	PGMV	30	Naidu _et al._, 1984b

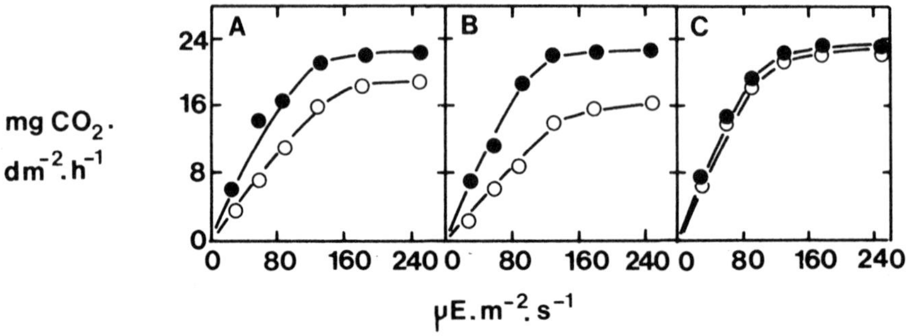

Fig. 5.1. Rate of carbon dioxide uptake by healthy (●) and PGMV-infected (○) peanut leaves as a function of light intensity at three infection stages: (A), stage 1, showing chlorotic spots; (B), stage 2, severe mosaic symptoms; (C), stage 3, recovery. Reproduced from Naidu _et al._ (1984b) by permission of the authors and Academic Press, London.

complete recovery of the leaf in biochemical as well as in visual terms.

A few authors, while measuring photosynthesis by gas exchange techniques, have studied further aspects of the gaseous diffusion process, in an attempt to discover which stage of photosynthetic CO_2 fixation was inhibited by infection. Thus in TAV-infected tomato, Hunter and Peat (1973) found no change in stomatal diffusion resistance, when the net photosynthetic rate was reduced by 40%. They suggested that changes in mesophyll resistance, inhibition of the CO_2 reduction pathway, or increased photorespiration, might be responsible. On the other hand, Lindsey and Gudanskas (1975) found that MDMV infection of maize caused a reduction in stomatal aperture, by inhibiting the active influx of K^+ ions to the guard cells. They suggested that this might have been responsible for the observed reductions in the rate of photosynthesis.

Hall and Loomis (1972a; 1972b) found that photorespiration of BYV-infected beet was 35% lower than in healthy plants; the lower net photosynthetic rate of infected plants could not, therefore, be explained by changes in photorespiration. In contrast, in the C_4 plant sugarcane, where photorespiration of healthy plants should be minimal (Ogren, 1984), Ghorpade and Joshi (1980) presented evidence that SCMV infection increased photorespiration. Synthesis of malate was reduced, and glycolate synthesis increased. These changes were accompanied by an overall reduction in the rate of carbon assimilation, and suggest that virus infection may have disrupted the measures which normally minimize photorespiration in C_4 plants.

An interesting report by Makovcova and Sindelar (1978) suggested that in tobacco systemically-infected with TMV, the CO_2-fixing activity of Rubisco was reduced by as much as 30% during the period of most active TMV accumulation. However, this was partly offset by increases in the activity of phosphoenolpyruvate carboxylase (PEP-Case), in both shoots and roots of infected plants. But Hall and Loomis (1972b) found no increase in PEP-Case activity in BYV-infected beet, and Mohammed (1973) reported a reduction in PEP-Case activity in TSWV-infected tobacco.

In contrast to reports of inhibitory effects of infection on photosynthesis, Doke and Hirai (1970) found evidence for a slight stimulation of CO_2 fixation in TMV-infected tobacco. However, they used a different assay, namely $^{14}CO_2$ fixation. They were probably measuring fixation into comparatively immobile forms such as protein and cell walls. This method gives information of a different type from the near-instantaneous gas exchange measurement methods used by most other workers. Fixation per unit chlorophyll was higher in the first day after infection, when there should have been little effect of the virus on chlorophyll content. Autoradiographic evidence was obtained for localized areas of higher fixation a few days later.

In SqMV-infected squash plants, Magyarosy et al. (1973) measured various parameters of photosynthesis, and concluded that infection had little effect. However, the experiments were done late in the infection, when the initially severe symptoms had moderated.

5.2.2. Chlorophyll content and photosystem activity

The energy pathway of photosynthesis involves the capture of a photon of light by the light-harvesting chlorophyll-protein complex, or associated accessory pigments. This causes excitation, or promotion of an electron to a higher energy orbital, and the energy is transferred to the reaction centres of either photosystem I (the P_{700}-chlorophyll-protein complex) or photosystem II (the presumed CP_a chlorophyll-protein complex reaction centre). The two photosystems operate in series, using two photons per equivalent moved through the electron transport chain, which involves many different electron carriers. The photosystems develop a chemical potential which is used to oxidize water and reduce ferredoxin, which reduces NADP. The chemical potential actually developed is higher than required, and some of the additional energy is used in phosphorylation of ADP to ATP (non-cyclic phosphorylation). In addition, photosystem I can operate in a cyclic manner to generate ATP.

Energy capture and conversion in photosynthesis is far more complex than can be conveyed in this brief summary. Recent reviews include

Butler (1978), Cogdell (1983) and Haehnel (1984). The purpose of the summary is to place the results of investigations with viruses in the overall context, and to emphasize that there are many stages at which infection could affect the capture and conversion of light energy.

Many investigators have measured changes in chlorophyll concentration after infection, and have tried to relate these to changes in photosynthetic activity. However, no overall single picture has emerged.

In their study of PGMV-infected peanuts at various stages of symptom expression (Fig. 5.1), Naidu *et al*. (1984b) found a good correlation between reduction in chlorophyll content and reduction in net photosynthesis at saturating light intensity. The data are summarized in Table 5.2. Chlorophyll *a* concentration tended to be more reduced by infection than chlorophyll *b*. Naidu *et al*. (1984b) concluded that the reduced rate of photosynthesis was due at least in part to the reduced chlorophyll levels. Further studies on chlorophyll-protein complexes by Naidu *et al*. (1984a) supported this conclusion: the chlorophyll-protein complex associated with the reaction centre of photosystem II was severely reduced in infected leaves, and there was also a reduction in light-harvesting chlorophyll-protein complexes. In contrast, the chlorophyll-protein complex associated with the reaction centre of photosystem I appeared to be increased in infected leaves. Further aspects of changes in these chlorophyll-protein complexes have been considered in Chapter 4.4.1.

However, infection of peanuts not only reduced chlorophyll and most chlorophyll-protein complexes, but also reduced the activities of photosystems I and II (measured per mg chlorophyll) by up to 15% and 43% respectively (Naidu *et al*., 1984b). By using a series of electron donors and acceptors, it was possible to examine in some detail which stages of electron transport were inhibited. Their conclusion was that plastoquinone, an intermediate carrier in the transport of electrons from photosystem II to photosystem I, was the probable site of inhibition. This was supported by fluorescence measurements indicating a reduced plastoquinone pool.

Table 5.2. Effects of peanut green mosaic virus infection on various photosynthetic parameters in peanut leaves.

Percentage change due to infection, in	Infection stage 1	2	3
Chlorophyll *a*	-19	-39	-10
Chlorophyll *b*	-14	-26	-9
Total chlorophyll	-17	-36	-9
Net photosynthesis	-18	-31	-2
Activities of:			
Photosystem I	-9	-15	+1
Photosystem II	-17	-43	-3
Non-cyclic phosphorylation	-26	-53	0
Cyclic phosphorylation	-5	-16	+1
Chlorophyll in chlorophyll-protein complexes:			
CPI	+15	+27	-1
CP_a	-15	-67	-4
$LHCP_3$	-7	-15	0

Chlorophyll and net photosynthesis were measured per unit leaf area, and photochemical activities and phosphorylation per unit chlorophyll. Infection stages were as described in the legend to Fig. 5.1. CPI is the chlorophyll-protein complex associated with photosystem I; CP_a the chlorophyll-protein complex associated with photosystem II, and $LHCP_3$ the light-harvesting chlorophyll *a/b*-protein complex. Derived from data in Naidu *et al*. (1984a; 1984b).

Naidu *et al.* (1984b) also showed that photophosphorylation was reduced in leaves with severe mosaic. Non-cyclic photophosphorylation was reduced by up to 53%, while cyclic photophosphorylation showed a reduction of only 16% at most. These results are consistent with the comparative insensitivity of photosystem I activity to infection, and its association with cyclic photophosphorylation. The sensitivity of non-cyclic photophosphorylation to infection could have been an indirect consequence of the effect of infection on plastoquinone.

In a later investigation of protein changes in infected leaves, Naidu *et al.* (1986) showed that the decreased capacity of photosystem II to evolve oxygen was associated with reduction in the quantities of five proteins from the chloroplast membranes. The function of these proteins and the mechanism controlling their amounts in virus-infected plants deserve further study.

It is clear from these studies that the effects of PGMV infection on the photobiology of peanut leaves are wide-ranging and complex, and that no single reaction can be identified as the stage at which infection inhibits photosynthesis.

Other host-virus combinations have been studied, but in much less detail. In tobacco infected with TRSV, Roberts and Corbett (1965) found that oxygen evolution per unit chlorophyll was significantly lower after infection, indicating that the observed chlorophyll loss of infected leaves was not the entire cause of reduced photosynthesis. Jensen (1968) found different results for two hosts infected by BYDV. Photosynthetic activity per unit chlorophyll was much higher in infected wheat than in healthy plants, and reduced to 50% of the healthy level in infected barley. Both hosts suffered similar, severe losses of total chlorophyll content as a result of infection.

In variegated leaves of *Tolmiea menziesii* (naturally infected with TBSV and CMV), Platt *et al.* (1979) found a lower chlorophyll content, but a higher rate of photosynthesis per unit chlorophyll, than in healthy leaves. Infection caused an increased chlorophyll *a* : chlorophyll *b* ratio, a decrease in the light harvesting chlorophyll-protein complex, but no change in the reaction centre chlorophyll-protein complex. Finally, in ToYMV-infected tomato, Leal and Lastra (1984) found that chlorophylls a and b were both reduced by about 50% by infection, and net photosynthesis decreased in proportion.

In sum, these results suggest that virus infections can reduce photosynthesis through an effect on chlorophyll concentration, but that this is far from being a simple linear relationship, and other factors can modify or even reverse the effects.

5.2.3. Chloroplast ultrastructure

The apparatus for capture of light energy in plants is highly structured, and contributes to function in different ways. The light reactions are associated with membrane structures, or lamellae, and the dark reactions of CO_2 fixation with the stroma or surrounding matrix. The lamellar system may be further sub-divided into grana lamellae (small thylakoids), where several layers of membrane are appressed to form stacks, and stroma lamellae (large thylakoids), which are unappressed. The light reactions of photosystem II occur in the grana lamellae, and those of photosystem I in the stroma lamellae. The whole chloroplast is bounded by a further double membrane structure, which is important for the pH gradient between cytosol and chloroplast, and for maintaining metabolic compartmentation.

Fig. 5.2 shows these ultrastructural features in a chloroplast from a healthy lettuce leaf (Tomlinson and Webb, 1978). In addition, the chloroplasts typically contained small starch grains, and a number of osmiophilic granules. Chloroplasts from plants infected with BWYV showed major alterations to chloroplast ultrastructure. At an early stage of infection, the lamellae of the grana had become more open and detailed structure was more clearly visible. The structure of the chloroplast was distended by vesicles containing large starch grains. At a later stage of infection, the lamellar structure of the grana had become completely disorganized, and the chloroplasts contained numerous large osmiophilic granules. These are thought to contain the lipid components derived from disorganized membranes. At this stage,

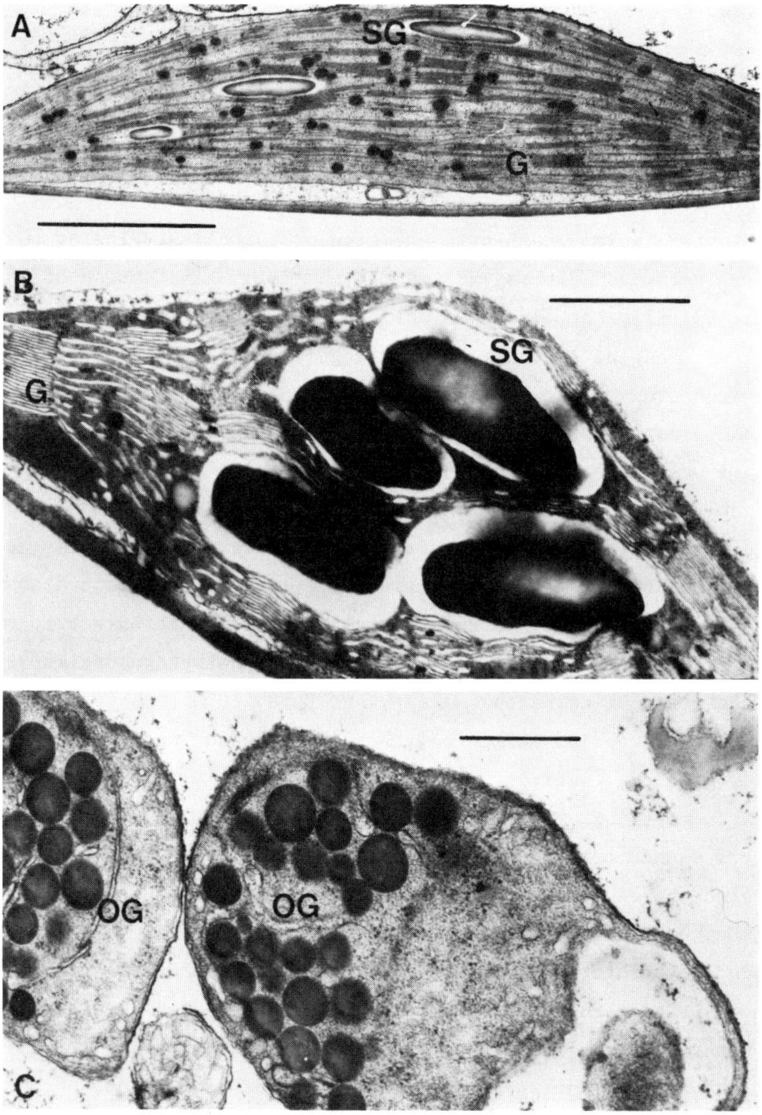

Fig. 5.2. Electron micrographs of chloroplasts from lettuce leaf mesophyll cells. (A) Healthy leaf; (B) and (C), infected with BWYV. (B) was sampled when the first signs of interveinal yellowing appeared, and (C) when the tissue had fully-developed symptoms and was white or pale yellow. SG, starch grain; OG, osmiophilic granule; G, grana. The bar indicates 1 µm. Electron micrographs by courtesy of M. J. W. Webb, National Vegetable Research Station.

infected tissue had lost almost all of its chlorophyll.

These changes during virus-induced yellowing were partly similar to those accompanying senescence. Tomlinson and Webb (1978) recorded partial disintegration of grana structure in healthy, yellowing leaves, but in contrast to virus infection, large starch grains did not develop.

A system which has been particularly well studied at both the cytological and ultrastructural levels is TYMV infection of Chinese cabbage. Different isolates of the virus cause mosaic of different colours, ranging from almost normal green to white, and the severity of structural changes in the chloroplasts is associated with this visible character (Matthews, 1973). Soon after infection, the virus induces small, double-membrane-bounded vesicles near the periphery of the chloroplasts (Hatta and Matthews, 1974). These are most probably the sites of viral RNA synthesis (Ralph et al., 1971; Mouches et al., 1974) (see Chapter 2.4.2). At this stage, the ultrastructural changes did not appear to have a deleterious effect on chloroplast function, as the Hill reaction, cyclic and non-cyclic photophosphorylation were all shown to be increased (Goffeau and Bove, 1965).

At later stages of infection, there were reductions in the size and number of grana (Ushiyama and Matthews, 1970) - especially with those isolates causing most loss of chlorophyll - and increases in the numbers of osmiophilic granules. Different isolates caused particular patterns of clumping or fragmentation of the chloroplasts (Chalcroft and Matthews, 1967a; Hatta and Matthews, 1974). Large vesicles also developed within the chloroplasts and caused severe distortion of chloroplast structure. This could include displacement of the chlorophyll-bearing membranes to one side to form a 'sickle'. Matthews and Sarkar (1976) showed that formation of sickled chloroplasts in TYMV-infected protoplasts was light-regulated and apparently dependent on photosynthesis, since it was completely prevented by 3-(3,4-dichlorophenyl)-1,1-dimethylurea (DCMU), an inhibitor of electron transport.

Appiano et al. (1978; 1981) compared changes in chloroplasts in Gomphrena globosa plants giving either a systemic or a local lesion

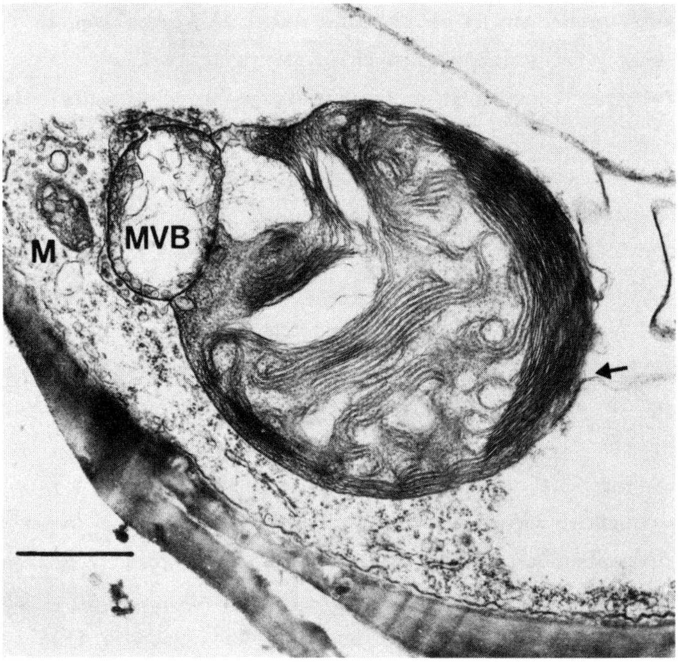

Fig. 5.3. Electron micrograph of a mesophyll cell chloroplast from a chlorotic region of a Gomphrena globosa leaf systemically infected with TBSV. The chloroplast is swollen, and the lamellae are disorganized. MVB indicates a multivesicular body and M a mitochondrion. The arrow indicates where the outer membrane of the chloroplast has ruptured. The bar indicates 1 μm. From Appiano et al. (1978) by permission of the authors and the Society for General Microbiology.

response to TBSV, although the systemically infected plants also had necrotic areas. In chlorotic patches of systemically-infected plants, the chloroplasts showed some swelling, with invaginations or vacuoles. There was some disorganization of the lamellar system, but not complete disruption. The most marked change was the formation of multivesicular bodies connected to the chloroplasts (Fig. 5.3). These were bounded by single membranes, and contained fibrillar material thought to be involved in virus multiplication. Some chloroplasts had ruptured outer membranes. Changes in chloroplasts near local lesions were generally similar to those in systemic infection, while chloroplasts in necrosing cells suffered complete collapse.

All or most of the changes in chloroplast ultrastructure associated with infection; that is vacuolation, large starch grains, disruption or loss of lamellae, increase in osmiophilic granules, and partial dissolution of the envelope, have been observed in a number of other host-virus combinations. These include TMV in tobacco (Shalla, 1964; Milne, 1966; Henry, 1983); RCMV in pea (Tomenius and Oxelfelt, 1982); PLRV in potato (Shepardson et al., 1980); BSMV in barley (Carroll, 1970) and avocado with the avocado sunblotch viroid (da Graca and Martin, 1981). There is therefore good evidence overall to associate inhibition of photosynthesis by viruses, with physical disruption of the chloroplasts. These cases refer, in general, to viruses which cause moderate to severe visible symptoms; there tends to be less physical damage to chloroplasts associated with mild visible symptoms. Magyarosy et al. (1973) suggested that SqMV had no effect on chloroplast ultrastructure in squash plants, but tissues were examined quite late after infection, when the initially severe symptoms had become progressively milder.

5.3. ASSIMILATE PARTITION AND CARBON BALANCE

5.3.1. The fate of photosynthetically-fixed carbon

A feature of many virus infections is alteration in the destinations of the products of photosynthetic CO_2 fixation. Several authors have reported reduced sugar levels, generally accompanied by an increased flow of assimilate to amino and organic acids. Bedbrook and Matthews (1972; 1973) found reduction in sugar phosphates and sugars in TYMV-infected Chinese cabbage, and increases in malic and aspartic acids. Gangulee et al. (1978) reported lower sugar concentrations in SBMV-infected cowpea, while Magyarosy et al. (1973) found a shift from sugar production to amino and organic acid production in SqMV-infected squash. The latter authors also reviewed a large number of earlier papers showing similar effects of other viruses.

Some authors have argued that the shift towards amino acid

synthesis reflects an increased requirement for these for synthesis of viral coat protein. Certainly, if sufficient synthesis of coat protein occurred to deplete the existing amino acid pools, there might be a feedback stimulation of further synthesis. On the other hand, it has been suggested that the reduction in sugar concentration might be due to feedback inhibition, as a result of the failure of infected plants to utilize the products of photosynthesis. The abundant ultrastructural evidence for the accumulation of large starch grains in the chloroplasts in many virus infections would support the idea that there is no shortage of photosynthate, although there might be a problem of translocation from the site of synthesis.

An interesting paper by Tu (1977) reported that 3'-5' cyclic AMP (cAMP) was reduced by CYMV infection of white clover. When healthy plants were treated with cAMP, there was an inverse correlation between the size of the starch grains in the chloroplasts, and the cAMP concentration applied. This led him to suggest that the accumulation of starch grains in chloroplasts of CYMV-diseased leaves might be a result of their reduced cAMP concentration. Although the occurrence of cAMP in plants now seems to be soundly established, there are still doubts about its role (Brown and Newton, 1981); phenolics and the natural degradation product 2'-3' cAMP can cause problems in assay.

In partial contrast to the reports of reduced synthesis of sugars, Jensen (1969a; 1972) showed that soluble carbohydrates were greatly increased in leaf blades of BYDV-infected wheat and barley, and that starch was also increased. He suggested that infection might interfere with the translocation of assimilate. BYDV is confined to the phloem, but does not appear to cause severe cytopathic effects there which might inhibit translocation (Jensen, 1969b).

Leal and Lastra (1984) showed that sugars and starch in ToYMV-infected tomato were higher than in healthy leaves at an early stage of infection, but lower at a later stage. Similar effects were reported in tomato infected with TBSV (Boninsegna and Sayavedra, 1978). Leal and Lastra (1984) also showed that late in infection, there was a change in the distribution of sugars and starch through the plant, compared with healthy controls. Lower leaves of infected plants had greater concentrations of sugars and starch than comparable healthy leaves, and upper leaves had less. A similar retention of sugars in lower leaves occurred in CuTV-infected tomato (Panopoulos et al., 1972). Leal and Lastra (1984) suggested that the accumulation of carbohydrate in lower leaves might reflect a failure of translocation: ToYMV multiplies mainly in the phloem and causes damage there (Lastra and Gil, 1981).

5.3.2. Where does the energy for virus multiplication come from?

Virus multiplication might utilize energy derived from respiration, from photosynthetic processes, or from both. TYMV synthesis in green tissue occurs only in the light, implying that the energy required is derived from photosynthetic ATP production. An intriguing experiment by Fernandez-Gonzalez et al. (1980) showed that TYMV synthesis in chlorophyll-less protoplasts, which were prepared from the hypocotyls of etiolated Chinese cabbage plants, could occur in either light or darkness. The suggestion was that energy was derived from respiration rather than the (non-existent) photosynthesis.

Kano (1985) studied the effects of light and inhibitors of photosynthesis and respiration on the multiplication of TMV in tobacco protoplasts. Multiplication was much higher in the light than in darkness. However, the fact that multiplication could occur in the dark suggested that respiration was a potential energy source. DCMU, an inhibitor of photosynthesis, prevented the stimulation of TMV multiplication in the light, and inhibitors of respiration did not block TMV multiplication in the light. This indicates that photosynthesis also could supply energy for virus multiplication.

5.4. CONCLUSION

The general impression which emerges is that virus infection increases catabolic processes such as respiration, and decreases anabolic processes such as photosynthesis. Mostly, it is difficult to say what the specific effect of the virus is on respiration or photosynthesis. Many investigators have looked at the overall processes, rather than trying to analyse the different stages. Furthermore, many or most of the phenomena which have been reported are likely to have been secondary effects, rather than primary sites of interaction with the virus.

For most effects described, it is not immediately apparent how the changes might benefit virus multiplication. Some of the effects must be secondary; inevitable consequences perhaps of events which are essential for virus replication. Others might express host defence reactions, or be related to damage-limitation and stress-response mechanisms. The almost universal tendency of viruses to damage chloroplast structure is particularly intriguing, because virus multiplication as such can be directly linked to the chloroplast only in a few cases.

What can be said with more certainty for photosynthesis, is that the various investigations provide a sound quantitative base of evidence for inhibition, which can be of use in analysing the effects of infection on growth and yield.

CHAPTER 6

The Biochemistry of Resistance to Viruses

6.1. BASIC CONCEPTS IN RESISTANCE STUDIES

6.1.1. Resistance in disease control

There are several methods for controlling the losses of yield and quality caused by virus infections of crops. Some aim to prevent infection, by improving crop hygiene and control of vectors such as aphids. Spread of virus through generations of the crop can be controlled by use of virus-free seed in the case of seed-transmitted viruses. Viruses can be eliminated from vegetatively-propagated crops such as potatoes by thermotherapy, chemotherapy and tissue culture (Walkey, 1980). At present, there are no chemical treatments, analogous to fungicides, which can be applied in the field to cure virus infections. There are chemicals which will prevent virus multiplication and symptom development in some infections, and with varying degrees of effectiveness, but they suffer from problems of cost and phytotoxicity (Fraser and Whenham, 1978a; Tomlinson, 1982; Cassells, 1983).

Resistance is the mechanism of choice for control of plant viruses, and resistance-breeding is a major objective for many crop species. Almost without exception, these types of resistance involve control by major genes, which are inherited in a simple Mendelian manner. However, the possibilities for control of virus diseases by resistance-type mechanisms are much wider. Increasing knowledge of the molecular interactions between plant and virus, and novel methods of handling genetic material, may allow better controls in the near future. This Chapter will consider the biochemistry of resistance in this wider context. The approach will be to consider various theoretical models, and then to review what is known of the biochemical mechanisms. The genetic background will be considered at each stage.

6.1.2. Resistance: definition of types and targets

The physiology and biochemistry of resistance mechanisms are poorly understood; what can be said is that the mechanisms are diverse. However, for reasons of convenience and labelling, as well as from the intellectual urge to classify, pathologists have long striven to define types and terms in resistance (e.g. Federation of British Plant Pathologists, 1973). This has led to confusion and semantic argument (Cooper and Jones, 1983; Tavantzis, 1984). It is therefore necessary to define certain key terms as they will be used in this Chapter, with the proviso that the terminology is one of convenience and is evolving. It is not absolute and definitive, but describes the best fit to current knowledge.

A plant species is a host for a particular virus when that virus multiplies in it and (normally) causes visible disease symptoms. The virus is pathogenic to the host.

Resistance is defined in the broadest possible sense in this Chapter, to mean any inhibition of virus multiplication, or of expression of its pathogenic effects. Virus multiplication is a process occurring in space and time, at the cell, plant and population levels. Resistance might operate at any of these levels.

We can envisage three types of resistance; these are based on concepts of underlying mechanisms, although a degree of overlap in the definitions is inevitable.

a) Non-host resistance is where an entire species is completely unaffected by a particular virus, in that no detectable multiplication occurs and no symptoms are formed, when the plants are inoculated. The virus in this case is non-pathogenic to the species.

b) Genetically-controlled resistance occurs when resistant

individuals within a host species contain a gene or genes conferring resistance to a particular virus normally affecting that species. Members of the species which do not contain the resistance gene are susceptible. This type of resistance is used by the plant breeder.

A virus isolate, normally pathogenic on the host species, which is inhibited by the resistance gene, is avirulent towards that gene. An isolate which is able to overcome a specific host resistance gene is virulent towards it. Note that virulence is therefore a qualifying attribute of the wider concept of pathogenicity. Confusion has arisen because the term 'virulence' has also been used to describe the severity of symptoms caused by virus isolates. The term 'aggressiveness' is preferred here. 'Aggressiveness' is also a qualifying attribute of pathogenicity, but does not necessarily have any functional relationship to resistance, or resistance-breaking behaviour.

Consideration of virus factors as well as host factors emphasizes the importance of studying the host-virus interaction, rather than either partner in isolation. A complex set of theories, with associated terminologies, has been developed to describe the types of interaction. These include the gene-for-gene model of interactions between resistance/susceptibility and virulence/avirulence, and the concepts of horizontal and vertical resistance (Flor, 1956; Person, 1959; Robinson, 1976; Vanderplank, 1984). The models have generally been developed from, and mostly applied to, studies of plant-fungus interactions. At this stage they have more to offer to the study of the genetics of plant-virus interactions than to the biochemistry, and are considered in that light by Fraser (1986). However, one point of particular relevance to mechanisms of resistance and virulence is the prediction of recognition events between host- and virus-specified molecules; this aspect is considered in section 6.1.5. An example of a complex gene-for-gene interaction between host and virus is considered in section 6.4.6. and Table 6.2.

c) Induced resistance occurs when plants of a host species have a type of resistance conferred upon them by a previous infection or other treatment. This type of resistance is not heritable, but must

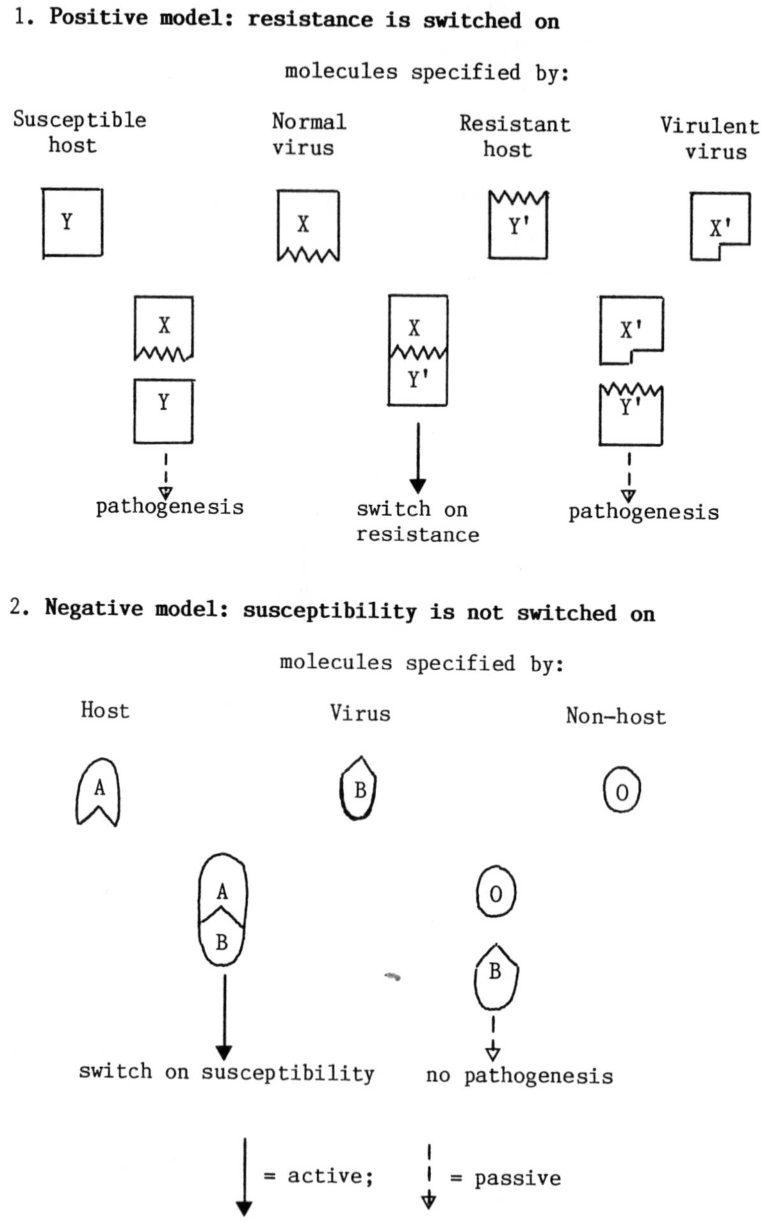
1. Positive model: resistance is switched on
molecules specified by:
Susceptible host
Normal virus
Resistant host
Virulent virus
Y
X
Y'
X'
X
Y
pathogenesis
X
Y'
switch on resistance
X'
Y'
pathogenesis
2. Negative model: susceptibility is not switched on
molecules specified by:
Host
Virus
Non-host
A
B
O
A
B
switch on susceptibility
O
B
no pathogenesis
= active;
= passive

Fig. 6.1. Continued opposite.

3. **The quantitative interaction model**

molecules specified by:

Normal virus	Virulent virus	Susceptible host	Resistant host
V	V'	H	H'

$$[V] + [H] \rightleftharpoons [VH] \longrightarrow \text{susceptibility}$$
$$[V] + [H'] \rightleftharpoons [VH'] \longrightarrow \text{resistance}$$
$$[V'] + [H] \rightleftharpoons [V'H] \longrightarrow \text{susceptibility}$$
$$[V'] + [H'] \rightleftharpoons [V'H'] \longrightarrow \text{susceptibility}$$

Fig. 6.1. Models for interactions between host- and virus-specified molecules which may determine susceptibility or resistance. In models 1 and 2, there is a critical 'recognition event' which can determine the outcome. Susceptibility and resistance are all-or-nothing consequences. In model 3, host- and virus-specified components react, but the equilibrium constant of the reaction depends on the nature of the components. Thus the concentration and properties of the host-virus compound will vary, and may induce a range of responses covering the whole spectrum from complete resistance to complete susceptibility.

be conferred afresh on each generation.

6.1.3. Positive and negative models of resistance

A plant might be resistant because it produces an inhibitor of virus multiplication. This is the positive model of resistance. Here, virulence on the part of the virus involves a failure to interact with the inhibitor, or an altered pattern of interaction such that pathogenesis can proceed (Fig. 6.1).

The alternative, negative model of resistance, is where the resistant plant lacks some component required for the viral replicative cycle. This could be a specific attachment or uncoating site, or a particular enzyme system such as an RNA polymerase or sub-unit on which the virus is totally dependent. Virulence in this model is the ability of the virus to complete its replicative cycle without the host component, or to mutate to use some other host component.

The models make certain predictions about the genetics of resistance and virulence, which are discussed in more detail by Fraser (1986) and Fraser and Gerwitz (1986).

6.1.4. The concept of resistance target

Resistance mechanisms, positive or negative, might interact with any of a number of stages of the virus replicative cycle; these stages can be divided into five groups. Broadly speaking, examples are known for all groups, and representative references are given for each. The biochemical mechanisms involved in some cases are discussed in more detail in sections 6.2-6.4.

Target Group 1: resistance to transmission. This includes resistance to vectors (Gibson and Pickett, 1983); vector non-preference (Pitrat and Lecoq, 1980), and resistance to seed transmission (Carroll et al., 1979). Clearly, some of the targets in this group are indirect, in that the resistance mechanism does not interact with the virus as such, but with another component of its replicative cycle.

Target Group 2: resistance to establishment of infection. This target is after the virus has been successfully delivered to the plant. The resistance of tobacco line T.I. 245 to infection by a number of viruses has been correlated with a reduction in the number of ectodesmata, which are thought to be the route of initial establishment of infection during mechanical inoculation (Thomas and Fulton, 1968). This is an example of a negative mechanism. This target area also includes the topic of virus-specific receptor sites, which is discussed in the context of non-host resistance.

Target Group 3: resistance to virus spread in the plant. This target includes a large number of mechanisms in different hosts, each of which restricts the virus to a small area around the site of infection (Holmes, 1938; Kim, 1970). Mostly, localization is accompanied by formation of necrotic lesions.

Target Group 4: non-localizing resistance to virus multiplication. The virus spreads systemically through the plant, but multiplication is inhibited, possibly at the levels of transcription, translation or assembly of progeny particles (Wyatt and Kuhn, 1979; Maule et al., 1980; Fraser and Loughlin, 1980).

Target Group 5: resistance to symptom formation. This target involves amelioration of the visible symptoms of virus infection such as mosaic and inhibition of growth. It does not necessarily imply inhibition of virus multiplication. Phenomena in this group are often referred to as 'tolerance' (Kooistra, 1968).

6.1.5. Recognition events between plants and viruses

In Target Groups 2 to 5, the virus and potential host come into contact, and at some stage in the process there is a point where the future of the association may be decided; a 'recognition event' in which host- and virus-specified factors interact to produce a 'go/no go' decision.

This interaction could involve either a switching-on, or a switching-off, of the further stages of pathogenesis (Fig. 6.1). In the former, a non-hosted virus lacks the specific molecule which interacts with the host to switch on phases of metabolism required for further development of pathogenesis. In the latter, recognition in a resistant plant could involve activation of a resistance mechanism such as localization; a susceptible member of the species lacks the specific recognition mechanism or means of responding to it.

Although recognition event-type models are popular when considering interactions between hosts and microbial pathogens, experimental evidence is still flimsy for virus infections. The models also imply that host-pathogen interaction is an all-or-nothing response. However, it remains possible that the outcome of a host-virus interaction might be decided by quantitative factors, especially in Target Groups 4 and 5.

6.2. BIOCHEMICAL MECHANISMS IN NON-HOST IMMUNITY

6.2.1. Possible genetic models

Non-host immunity mechanisms might represent limit cases (i.e. completely effective) of the positive and negative resistance models shown in Fig. 6.1. Holmes (1955) suggested that non-host immunity might result from a collection of as many as 20 to 40 individual resistance genes, each partly effective against the virus. In support of this positive model, he cited the work of Wade and Zaumeyer (1940) on resistance to AMV in Phaseolus vulgaris. The reaction of 'susceptible' plants to infection was the formation of small necrotic lesions, in itself a resistance response. Presence of either of two further genes conferred complete immunity. However, this work only provided evidence for additive effects of genes, but not for the involvement of a large number of loci. The conclusions cannot necessarily be extrapolated to completely immune (non-host) species.

In a classical series of papers, Bald and Tinsley (1967a; 1967b; 1967c) developed a quasi-genetic model to explain host range and non-host immunity. In contrast to the Holmes model, theirs is negative, in that non-host immunity results from a lack of some factor in the host or virus. The model proposes that the host contains a number of susceptibility genes, and the virus a number of pathogenicity genes. A successful infection is established only when the particular sets of susceptibility and pathogenicity genes are complementary. Thus a virus containing genes abc can successfully infect a host containing genes ABC, but not one with genes XYZ or ABD. It should be noted in this context that the definition of 'gene' for the virus has to be fairly flexible, in view of the limited coding capacity of the complete viral genome. More than one determinant of pathogenicity could reside in each cistron, and the cistron could code for some other function as well.

The quasi-genetic model can offer an explanation for the phenomena known as host range congruence and containment. Congruence is where apparently unrelated viruses show a tendency to infect the same host species. Containment is an extreme case of this, where virus A will infect all species infected by virus B, while virus B will infect some of the species infected by virus A but none which are not infected by virus A. The implications are that in congruence, the two viruses have certain pathogenicity factors in common. In containment, virus B has some, but not all, of the pathogenicity factors of virus A, and no others.

6.2.2. Are there virus–specific receptors?

A receptor is a site where the virus attaches to the host cell, and may also be involved in penetration and uncoating. Virus–specific receptors on animal and bacterial cell surfaces are comparatively well understood (see Chapter 2.2.1), but the picture is much less clear for plant cells. The ways in which plant cells become infected pose problems: mechanically–transmitted viruses normally require wounding of the recipient cell, to allow penetration by the virus. Vector–transmitted viruses are delivered directly into the host cells. In both cases, external attachment sites are somewhat unnecessary, although internal recognition sites could still be important.

Most studies of putative receptor sites have presupposed that the surfaces of coat protein exposed on the outside of the virus particle are likely to be important in recognition between virus and receptor. One approach which has been used is to reconstitute virus particles from RNA and protein, so that the RNA of one virus is encapsidated in the protein of another. Sometimes, such 'hybrid' or heterologously encapsidated particles can have as high an infectivity as the original viruses (Hiebert et al., 1968), although in other cases infectivity may be poor (Atabekov et al., 1970).

Hiebert et al. (1968) studied three bromoviruses, BMV, CCMV and BBMV (Fig. 6.2). Chenopodium hybridum and soybean are immune to BBMV and susceptible to CCMV. CCMV RNA encapsidated in BBMV coat protein was able to infect both hosts successfully. This suggests that CCMV does not rely on CCMV–specific recognition sites for infection. Furthermore, both hosts were able to uncoat the CCMV RNA successfully, even though it was encapsidated in the coat protein of

Fig. 6.2. Possible mechanisms of host-range control based on specific receptor sites. Three bromoviruses, BBMV, CCMV and BMV, were disassembled and mixedly reassembled particles were prepared from the proteins (represented by closed symbols) and RNAs (represented by open symbols). (————) shows successful establishment of infection; (—✗—) that the species is a non-host for that virus or mixedly reconstituted particle. (------) shows the sources of components for mixed reconstitutions. Based on data in Hiebert et al. (1968) and re-drawn from Fraser (1985c).

a non-hosted virus. This would tend to rule out any virus-specific uncoating mechanism. Other examples of successful infection by viral RNA encapsidated in the coat protein of a non-hosted virus are given by Atabekov et al. (1970). Conversely, Shaw (1969) reported that TMV RNA was uncoated in plants which are not hosts for this virus, and which did not support its replication.

The opposite conclusion about virus-specific receptors and uncoating may be drawn from the experiments of Atabekov (1975). He showed that BMV RNA, if encapsidated in TMV coat protein, failed to infect barley, although the heterologously-encapsidated particles were infectious to other species. The barley cultivar used was said to be 'immune' to TMV, although some other cultivars support a low

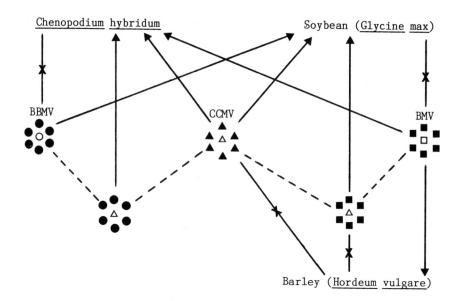

level of multiplication of this virus (Dodds and Hamilton, 1974). One interpretation of Atabekov's data, therefore, is that the particles reconstituted with TMV coat protein failed to interact with a BMV-specific attachment site, or failed to uncoat. However, the latter explanation could be the result of trivial causes, not directly related to host range control.

Another type of experimental approach has been to encapsidate the RNA of a non-hosted virus in the coat protein of a hosted virus, to see if this allowed it to infect the (normally) non-host species. Hiebert et al. (1968) showed that CCMV RNA encapsidated in BMV would not infect barley, a non-host for CCMV (Fig. 6.2). This suggests that interaction or failure of interaction with a putative specific receptor site cannot be solely responsible for non-host immunity. However, there is also a problem in interpretation of this and other experiments involving mixed reconstitution. The heterologously-encapsidated particle might indeed behave as predicted from the specific receptor site hypothesis, but only in the initially infected cell. Further virus spread would involve progeny virus particles which would be encapsidated in the homologous protein, having been produced by translation from the RNA component. Thus the primary infection and subsequent spread might be subject to different constraints.

Novikov and Atabekov (1970) used a different type of approach to the study of receptors. They examined competition between whole virus particles, and homologous or heterologous coat proteins. For PVX, homologous coat protein did not interfere with infection, while with BSMV and TMV, homologous coat protein interfered but heterologous did not. These data are consistent with, but do not prove the existence of, specific attachment sites for BSMV and TMV. However, very high concentrations of coat protein were required for competition. A meaningful experiment would have been to use coat proteins assembled into virion-like structures, where possible, as competitor.

An interesting approach to the possible role of cell-surface features in non-host immunity has been to seek to remove or bypass them, by attempts to infect protoplasts of non-host species. Furusawa and Okuno (1978) showed that BMV could infect protoplasts of Japanese radish, but could not infect whole leaves. The tobacco strain of TMV caused a virtually undetectable infection of cowpea leaves, but replicated well in protoplasts (Huber et al., 1977; 1981). These experiments might suggest that non-host immunity of these species takes the form of some barrier to infection at the cell wall level. However, an alternative explanation is possible. Sulzinski and Zaitlin (1982) showed that cowpea leaves inoculated with the tobacco strain of TMV contained a very small number of infected cells, but appeared not to support multiplication of the virus because movement from cell-to-cell was inhibited.

There are also reports of failure of viruses to multiply in protoplasts of non-host species (Motoyoshi et al., 1974).

In summary, the evidence for involvement of virus-specific receptor sites in non-host immunity and host range control is patchy, and many of the experiments could have alternative explanations.

6.2.3. Possible recognition events at transcription and translation

Although some of the more complex plant viruses, such as rhabdoviruses, contain a transcriptase activity as part of the virion (Toriyama and Peters, 1980), there is evidence that others rely at least in part on host-coded enzymes for transcription (see Chapter 2). Clearly, any replicase or RNA-dependent RNA polymerase which depends on sub-units coded by both host and viral genomes offers opportunities for recognition and host range control. Until recently, the evidence argued against virus-coded sub-units in RNA-dependent RNA polymerases of infected plants (Romaine and Zaitlin, 1978; Gordon et al., 1982). However, the relationship between the highly-purified polymerases studied in vitro and the true replicase activity in vivo is not clear. Recent work on the putative replicase of TYMV does suggest that it contains both host- and virus-coded sub-units (Mouches et al., 1984). Non-host immunity operating at the replicase level is therefore possible, but to date there is no direct evidence for it.

Non-host immunity could also operate at the translational level if there is a recognition event, or failure of recognition, between viral RNA and the ribosomes of the plant cell. There is no evidence yet for specificity at this level. The ability to translate viral RNAs in a wide range of cell-free in vitro translation systems would tend to argue against recognition at this level, although it must again be remembered that these systems may not truly reflect what is happening in vivo. Kiho et al. (1972) found that TMV RNA was found in polyribosomes as much in host plants as in those which were virtually non-hosts, suggesting that non-host immunity does not act at this stage.

Maekawa et al. (1981) found that BMV would multiply poorly in protoplasts from the non-hosts radish and turnip, but that actinomycin-D or ultraviolet light treatment enhanced multiplication. Both treatments tend to inhibit host protein synthesis rather than that of the virus; the suggestion was that BMV synthesis in these plants would normally be suppressed by products of host transcription and translation.

6.2.4. Complementation experiments

There is very limited evidence that a 'helper' virus can assist a second virus to replicate in a plant which is normally a non-host for the second virus. In the Bald and Tinsley (1967a) model, this could be interpreted as the pathogenicity factors of the helper virus making good any deficiencies in those of the normally non-hosted virus. However, the evidence is indirect, and does not apply to a true non-host situation. Thus TMV in barley is normally restricted to the inoculated leaf, and does not spread systemically (Hamilton and Dodds, 1970; Hamilton and Nichols, 1977). Simultaneous infection with BSMV or BMV, however, allowed high levels of systemic TMV multiplication. There was evidence for some encapsidation of TMV RNA in BSMV coat protein, but not in BMV coat protein (Dodds and Hamilton, 1974). Heterologous encapsidation in the protein of a hosted virus cannot therefore have been the explanation of the enhanced systemic spread of TMV. Hamilton and Nichols (1977) also found that BMV did not allow several other viruses, which do not

Fig. 6.3. Glandular hairs on the leaf surface of <u>Solanum</u> <u>berthaultii</u>. The scanning electron micrograph shows two types of hairs: Type A with four lobes, and the longer Type B ending in a single sticky droplet. The bar shows 100 μm. Photograph by Dr R. A. Gibson, Rothamsted Experimental Station.

replicate alone in barley, to do so in mixed infections.

6.3. GENETICALLY CONTROLLED RESISTANCE TO TRANSMISSION, AND DISEASE AVOIDANCE

Host features such as colour and hairiness, which may affect whether flying vectors settle and thus confer some resistance to spread of the virus in the population, have been reviewed by Gibson and Plumb (1977). As the mechanisms of resistance are indirect, and there is no detailed biochemical knowledge of them, they are not considered in detail here.

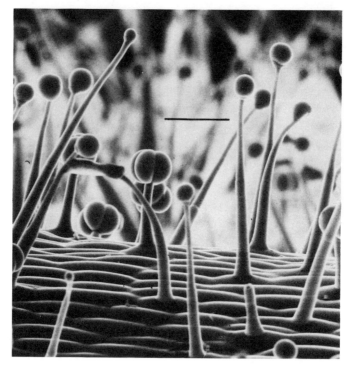

One example which is better understood in biochemical terms concerns the wild potato Solanum berthaultii. Leaves bear two types of hair, which entangle insects (Fig. 6.3). One type (B) also exudes the aphid alarm pheromone (E)-β-farnesene. Aphids produce this, for example when attacked by a predator, and other aphids respond by moving away (Nault and Montgomery, 1977). The amount of alarm pheromone produced by S. berthaultii is enough to cause aphids to avoid the plant (Gibson and Pickett, 1983) and to reduce acquisition and transmission of BYV and PVY (Dawson et al., 1982).

Another mechanism which may influence transmission involves substances dissolved out of the plant cuticle during salivation and their detection by chemoreceptors on the insect's labium (Tarn and Adams, 1982). For example, the aphid Myzus persicae will salivate on certain potato lines which it finds unacceptable as hosts, but fails to complete penetration of the phloem by the stylet (Adams and Wade, 1976). This would prevent acquisition and transmission of viruses. However, the biochemical basis does not seem to have been explored.

6.4. GENETICALLY CONTROLLED RESISTANCE: MECHANISMS OPERATING WITHIN THE PLANT

6.4.1. An overview of the genetic controls

Resistance in most host-virus combinations analyzed seems to be very simple at the genetic level. A recent survey of 63 randomly-chosen examples showed that over 90% of the mechanisms appeared to be under monogenic control (Fraser, 1986, and Table 6.1). This tends to suggest that the biochemical mechanisms of resistance should also be comparatively simple in terms of numbers of components involved in the primary response.

Most of our knowledge of the genetics of resistance is based on studies of cultivated crops. These are the products of extended breeding programmes, which may have led to a simplification of the genetic basis of resistance. Resistance genetics in wild species might be more complex.

Resistance in a large number of host-virus combinations operates by

Table 6.1. Genetics of resistance to viruses in crop species, and some features of resistance gene action and virulence. Derived from data in Fraser (1986).

GENETIC BASIS:	Number of host-virus combinations	
Single dominant gene	29	
Incompletely dominant (gene-dosage dependent)	10	
Apparently recessive	11	
Sub-total: monogenic		50
Possibly oligogenic	5	
Monogenic, with possible modifier genes or effects of host genetic background	8	
Sub-total: oligogenic (?)		13
Total number of host-virus combinations in sample		63

LOCALIZATION:	Immune	Yes	Partial	No	Not known	Total
Dominant alleles	0	19	0	2	8	29
Incompletely dominant	0	0	4	8	0	12
Apparently recessive	5	1(?)	1	2	4	13

'Immune' = no virus detectable. 'Yes' = normally involving lesion formation. 'No' = resistance permitting some systemic spread. 'Not known' = not tested, or not reported in the literature.

TEMPERATURE RESPONSE:	ts	tr	Not known	Total
Dominant alleles	7	2	20	29
Incompletely dominant	1	1(?)	10	12
Apparently recessive	2	2	9	13

'ts' = temperature sensitive. 'tr' = temperature resistant.

VIRULENT ISOLATES REPORTED:	Yes	No	Not known	Total
Dominant alleles	16	1	12	29
Incompletely dominant	8	3	1	12
Apparently recessive	4	1	8	13

preventing spread of the virus from the site of infection. Generally, the localization response is controlled by a single dominant gene, or rarely by an incompletely dominant gene. Localization controlled by recessive genes is extremely rare or non-existent (Fraser, 1985b; Table 6.1). Recessive and incompletely dominant genes either allow limited systemic spread of virus, or give apparently complete immunity. Cases of completely recessive behaviour appear to be very rare; many reported so-called recessive genes have turned out to show gene-dosage dependence when the phenotype of the heterozygote was examined. The rarity of fully recessive resistance suggests that the negative model for resistance gene action is uncommon.

The apparent genetic distinction - and there are some exceptions - between localizing and non-localizing resistance mechanisms may be rationalized by regarding the localization response as a qualitative, or all-or-nothing event, which should operate in heterozygous as well as in homozygous resistant plants. This would give phenotypic dominance. Resistance permitting some systemic spread is more likely to involve quantitative interactions, and then gene-dosage effects would be likely.

6.4.2. Virus localizing mechanisms

The localization of virus is normally accompanied by a hypersensitive response, leading to a necrotic local lesion, or less commonly by formation of a chlorotic lesion (Table 6.1). In a very few cases the response is symptomless or microscopic, and confined to one or a small number of cells (Sulzinski and Zaitlin, 1982).

Most evidence suggests that necrosis is not directly involved in localization. Virus can frequently be detected in tissue surrounding the necrotic area (Hayashi and Matsui, 1965; da Graca and Martin, 1976; Konate et al., 1983). Also, the size of the necrotic region can be reduced without apparently reducing virus spread or multiplication, for example by treatment with cytokinins (Balazs et al., 1976) and antibiotics (Takusari and Takahashi, 1979).

However, the suggestion that necrosis is not involved in localization leaves us with an unsolved problem: why then is it so often associated with localization resistance? One possibility is that it might prevent the localized virus serving as inoculum for further infection. The alternative explanation is that necrosis is normally an inevitable secondary consequence of the changes which primarily lead to localization.

Localization appears to be an induced response in most infections, in that several hundred cells often become infected before it becomes fully operative. Thus the early rates of TMV multiplication in susceptible and hypersensitive tobacco cultivars are similar (Taniguchi, 1963; Otsuki et al., 1972). The fact that many viruses which elicit a local lesion response can multiply well in protoplasts from resistant leaves (Otsuki et al., 1972; Motoyoshi and Oshima, 1979) is also cited as evidence for the induced nature of localization. Only one cycle of replication occurs in protoplasts, and this gives less time for induction of the resistance mechanism than in leaves where successive cycles occur. However, there are qualifications to these arguments. Virus multiplication in isolated protoplasts can be less than in leaf cells (Loebenstein et al., 1980), and the conditions of culture might prevent the accumulation of sufficient concentrations of those substances which trigger necrosis. There is some evidence for accumulation of a protoplast-damaging factor in the culture medium, for TMV-infected protoplasts prepared from a hypersensitive tobacco cultivar (Hooley and McCarthy, 1980).

There is also at least one example where the localization mechanism in the leaf is constitutive; TMV in cowpea is restricted to the primarily inoculated cells (Sulzinski and Zaitlin, 1982). However, this probably involves a mechanism of resistance which is quite different from that involved in necrotic lesions formation.

Virus localizing mechanisms have generally turned out to be temperature-sensitive, where this has been examined (Table 6.1), with systemic spread of virus and failure of necrosis occurring above about 30°C. The temperature profile of the transition from the localizing to the systemic response seems to be steep (e.g. Fraser, 1983). This might suggest an abrupt event such as a phase-transition in membrane structure, rather than progressive thermal denaturation of an inhibitor. Temperature-sensitive localizing resistance is also commonly controlled by phenotypically dominant genes (Table 6.1). This is consistent with a positive resistance mechanism involving an all-or-nothing recognition event (Fig. 6.1).

A fuller account of the biological aspects of localization and lesion formation is given by Fraser (1985d).

6.4.3. Metabolic changes accompanying localization

Dramatic changes in several areas of metabolism accompany localization; Fig. 6.4 summarizes the timing. A general increase in respiration in tissue around lesions was noted in Chapter 5, and serves as an indicator of increased metabolic activity. The problem is to separate changes which may be directly involved with resistance, from those which are secondary effects, or consequences of necrosis. Some use has been made of inhibitors to test causal relationships. In this section, most weight will be given to metabolic events which may be related to resistance.

Changes in transcription and translation. Inhibitors of transcription and translation have been used in attempts to discover whether localization and necrosis depend on activation of host genes. Actinomycin-D and ultraviolet light partially suppressed localization of TMV in tobacco and bean (Loebenstein *et al.*, 1969; Wu and Dimitman, 1970), in that they caused larger lesions with more infectivity. However, inhibitor experiments are difficult to interpret, as the inhibitor may directly affect virus multiplication (Loebenstein *et al.*, 1970). Inhibitors might also operate at the level of general phytotoxicity rather than by an effect on a specific gene. The more specific approach has been to examine changes in individual enzymes which might operate in resistance.

Phenolic compounds and oxidative enzymes. Phenolic compounds and derivatives, such as chlorogenic and isochlorogenic acids and scopoletin, accumulate during necrogenesis, and have been implicated

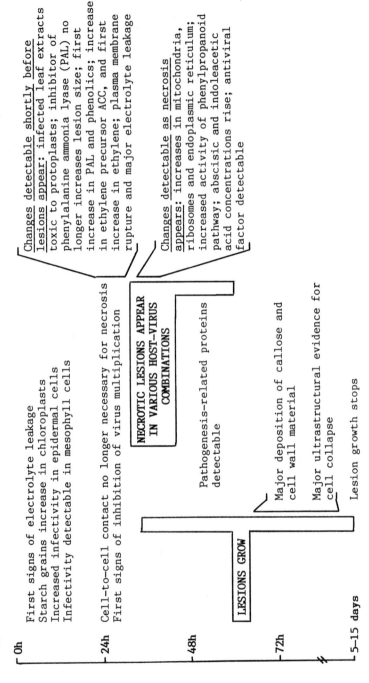

Fig. 6.4. The relative timings of some metabolic events during localization of virus and formation of necrotic lesions. References are given in the text.

in that process (Cruikshank and Perrin, 1964; Tanguy and Martin, 1972). Time course experiments suggested that the increases occurred only after necrosis became visible, but other experiments using more sensitive methods suggested that concentrations and rates of synthesis of chlorogenic acid and scopoletin were detectably higher before lesions appeared (Fritig et al., 1972).

The increased activities of peroxidase and polyphenoloxidase during necrogenesis were discussed in Chapter 4. It has been suggested that accumulation of their products, toxic quinones, might reduce the rate of virus multiplication and lesion spread (Van Loon and Geelen, 1971; Tanguy and Martin, 1972). However, time course experiments indicated that these events were too late to be major controls of localization (Fritig et al., 1972; Weststeijn, 1976).

Phenylpropanoid metabolism. The phenylpropanoid pathway serves as a source of precursors for lignin synthesis. Examination of the ultrastructure of areas around lesions suggests that lignification and wall modifications accompany necrogenesis (e.g. Appiano et al., 1977; Faulkner and Kimmins, 1975; Favali et al., 1978). The evidence for involvement of lignification and other wall changes in localization is fairly comprehensive, but several authors have argued against a primary role (e.g. Russo et al., 1981; Appiano et al., 1977; Pennazio et al., 1978).

The sequential activation of the enzymes of the phenylpropanoid pathway during localization has been discussed in Chapter 4.4.5. Legrande et al. (1978) suggested that the induced thickening of cell walls could inhibit virus spread at the lesion edge. Phenylalanine ammonia lyase (PAL), the first enzyme of the pathway, is inhibited by αamino oxyacetate (AOA). Massala et al. (1980) showed that AOA increased the size of TMV lesions by up to four times if applied before lesion appearance (Fig. 6.5). It did not suppress the normal increases in activities of the enzymes of phenylpropanoid metabolism after infection, but did inhibit pathway flux. This is demonstrated in Fig. 6.5 as a decrease in accumulation of chlorogenic acids. These experiments therefore provided evidence for a role of the

124

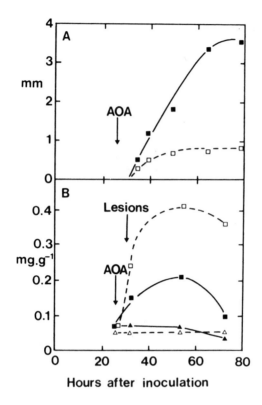

Fig. 6.5. Effects of α-amino oxyacetate (AOA) on lesion growth in
TMV-infected <u>Nicotiana</u> <u>tabacum</u>, and on biosynthesis of phenolic
compounds induced by infection. (A). Lesion growth (diameter) in
control leaves (□), and in leaves treated with AOA from 25 h after
inoculation (■). (B). Changes in leaf concentrations of chlorogenic
acids. (△), healthy leaf. (▲), healthy leaf treated with AOA from
25 h. (□), infected leaf, untreated. (■), infected leaf treated
with AOA from 25 h. Reproduced from Massala <u>et al</u>., 1980, by
permission of the authors and Academic Press.

activated pathway in limiting virus spread, or at least in limiting
necrosis. The experiments also showed that the large lesions on
AOA-treated leaves contained less virus per unit necrotic area than
lesions on control leaves; this result is difficult to explain.

Subsequent studies (R. Massala, M. Legrand and B. Fritig, personal
communication) used α-aminooxy-β-phenylpropionic acid (AOPP). This is

the true aminooxy analogue of phenylalanine and a more potent and specific inhibitor of PAL than is AOA. AOPP inhibited the TMV-induced accumulation of the soluble phenylpropanoids scopoletin and chlorogenic acid very strongly, but surprisingly, had less effect on lesion size than AOA. Lignin accumulation was inhibited more strongly by AOA than by AOPP. This led Fritig and his co-workers to suggest that the products of the phenylpropanoid pathway which are important in resistance are the insoluble lignin polymers, and that AOA may affect lesion growth by acting at a stage of the pathway later than PAL.

Martin-Tanguy *et al.* (1976) found that TMV-infection of hypersensitive tobacco led to an increase in aromatic amides, which also require precursors from the phenylpropanoid pathway. Some of these amides inhibited TMV multiplication.

Pathogenesis-related (PR) proteins. Although it has been suggested that PR proteins are involved in the induced resistance of hypersensitively-reacting plants to a second inoculation (Kassanis *et al.*, 1974; Chapter 6.5), there have been surprisingly few experiments testing whether they play any part in resistance to the first inoculation. PR proteins in tobacco become detectable 3 to 4 days after inoculation with TMV (Van Loon and Van Kammen, 1970), at about the time of lesion appearance. By this time, localization of the virus is probably well established (Takahashi, 1973). The accumulation of PR proteins is probably a response to the burst of ethylene synthesis at the time of lesion appearance (Van Loon, 1983a). In TMV-infected hypersensitive plants grown at a low temperature after infection, to allow localization and PR protein accumulation, and then shifted to a higher temperature, localization broke down and the virus spread systemically despite the continued presence of PR proteins (Van Loon, 1975). In sum, therefore, the evidence suggests that PR proteins are not directly involved in the localization of the primary infection.

Plant growth regulators. Application of exogenous plant growth regulators to plants displaying the hypersensitive response causes a bewildering range of effects on lesion development and localization. The data, reviewed by Fraser and Whenham (1982), give little insight into how localization occurs, and the role - if any - of endogenous growth regulator metabolism in localization.

Somewhat rarer studies of the metabolism of endogenous growth regulators have shown increases during localization, for example of auxin (Van Loon and Berbee, 1978); abscisic acid (Whenham and Fraser, 1981), and ethylene (Gaborjanyi *et al*., 1971; Pritchard and Ross, 1975; De Laat *et al*., 1981). The evidence linking auxin and ABA to control of localization is poor, and for ABA the major increases occur after necrosis. For ethylene, the major peak of synthesis is after necrosis, but this is preceded by an increase in the concentration of the ethylene precursor 1-amino cyclopropane-1-carboxylic acid (ACC) before lesions become visible (De Laat and Van Loon, 1982). The balance of the evidence argues against any controlling role for ethylene in the onset of necrosis and localization, although suggesting a major part in control of related aspects of metabolism such as the induction of pathogenesis-related proteins (Van Loon, 1977; 1983b).

Membrane and permeability changes. Increased permeability of the cell membrane, measured as increased leakage of electrolytes from cells, was found in TMV-infected tobacco (Weststeijn, 1978) and TRV-infected cowpea (Pennazio and Sapetti, 1982). The leakage occurred in advance of necrosis, at about the time when the pathway for ethylene synthesis was becoming activated. The two events may be linked, as ethylene synthesis is thought to occur at the plasmalemma (Lieberman, 1979). Using a different experimental procedure, Ruzicska *et al*. (1983) found that a rapid increase in the saturation of phospholipid-bound fatty acids occurred before the increase in electrolyte leakage. The earliest membrane change detectable after infection was an increase in activity of a membrane-bound ATPase (Kasamo and Shimomura, 1978), but this was not specific for localization or necrosis.

Pennazio and Sapetti (1982) concluded that increased permeability is not involved in localization, but is a non-specific accompaniment of necrosis. This is supported by experiments with Nicotiana glutinosa showing that membrane-damaging agents caused lesion formation, which could be prevented by membrane stabilizers such as polyamines and calcium (Martin and Martin-Tanguy, 1981; Ohashi and Shimomura, 1982).

6.4.4. Antiviral factors

The experiments on metabolic changes accompanying localization have been largely descriptive, and in some cases have not been based on any well-developed hypothesis for involvement in resistance. In contrast, the aim in another type of experiment has been to isolate and characterize molecules with a demonstrable antiviral activity. There have been two main types of approach: to isolate agents from one species, either infected or not, which will protect plants of the same or a different species from challenge infection; and to isolate compounds thought to be involved in the primary localization response of the inoculated plant.

Assays of antiviral activities of both types of agent have mostly involved measurement of the reduction in numbers of lesions formed when the putative antiviral compound plus virus are applied to a test host. This is likely to reflect effects on the establishment of infection, or on host susceptibility to infection. Fewer experiments have sought to measure effects on virus multiplication as such. Experiments with the latter aim also need to demonstrate some measure of specificity of action, and must exclude the possibility that the putative inhibitor is merely some non-specific metabolic poison.

A third model which has been proposed is that the antiviral substance may not inhibit virus infection or multiplication directly, but may induce some antiviral state in host cells, perhaps also acting as a messenger to cells distant from the site of infection.

Inhibitors effective in a second plant species. Endogenous, or pre-formed inhibitors of plant viruses have been isolated from healthy plants of a number of (donor) species, including pokeweed (<u>Phytolacca</u> <u>americana</u>) (Owens <u>et</u> <u>al</u>., 1973); carnation (<u>Dianthus</u> <u>caryophyllus</u>) (Fantes and O'Neill, 1964); <u>Boerhaavia</u> <u>diffusa</u> (Verma and Awasthi, 1980), and <u>Bougainvillea</u> <u>spectabilis</u> (Verma and Dwivedi, 1984). These, when mixed with virus, or when applied to the second (test) species before inoculation, will reduce infection and virus multiplication. Generally, they are active against a spectrum of viruses, and in hypersensitive or systemically-infected test species. Further examples of donor species are given by Chessin (1983).

The inhibitors which have been characterized appear mostly to be proteins, frequently glycoproteins (Irvin <u>et</u> <u>al</u>., 1980; Verma and Awasthi, 1980), although other reports include RNA (Kimmins, 1969) and low molecular weight phenolics (Schuster and Wetzler, 1982). Grasso and Shepherd (1978) and Chessin (1983) have drawn attention to the richness of the order Centrospermae as a source of potent antiviral proteins; the former authors showed that proteinaceous inhibitors of virus infection from some species were serologically and otherwise related. However, they also found inhibitors of viral infectivity from other, taxonomically unrelated species.

Inhibitors of virus infectivity have also been reported to be stimulated by localized (Loebenstein and Ross, 1963; Kimmins, 1969) or systemic infection of the donor species (Chadha and MacNeill, 1969; Nienhaus and Janicka-Czarnecka, 1981). Again, there appears to be little specificity; infection with one virus can stimulate an inhibitor of infectivity of numerous other viruses in the test species.

Several of the inhibitors have been reported to induce systemically-effective resistance in the test species. Induction of resistance has been shown to be prevented by treatment with actinomycin-D (Verma and Baranwal, 1983; Verma <u>et</u> <u>al</u>., 1984), suggesting that the generation of resistance requires transcription of host genes. Verma and Awasthi (1980) showed that treatment of test plants of several species with the inhibitor from <u>Boerhaavia</u> <u>diffusa</u> roots caused production of antiviral agents, thought to be proteinaceous. The antiviral agents were produced from 2 to 48 h after treatment, and were active in other test species as well as in the one in which they had been produced. The antiviral agents reduced infectivity of viruses and virus multiplication in leaves. Verma and Dwivedi (1984) reported induction of virus-inhibiting agents with similar properties after treatment of a number of test species with an extract of Bougainvillea spectabilis leaves. With most such inducers and, where appropriate, secondary inhibitors, there remains a requirement to demonstrate the modes of action, and to exclude trivial explanations.

The inhibitors from Phytolacca and Dianthus are amongst the best understood in terms both of composition and mode of antiviral action. Stirpe et al. (1981) purified to homogeneity two proteins from Dianthus caryophyllus, of molecular masses 29.5 kDa and 31.7 kDa. Both proteins, called dianthins, contained mannose, and both inhibited protein synthesis by a rabbit reticulocyte lysate, at concentrations of less than 10 ng/ml, i.e. they were effective at less than an equimolecular ratio. Each molecule of dianthin was therefore inactivating more than one ribosome, suggesting a catalytic or enzymatic mode of action. A much higher concentration (100 µg/ml) was required for partial inhibition of protein synthesis in mammalian cells in tissue culture: this might have reflected failure of uptake by the cells. Strong inhibition of local lesion formation by TMV on Nicotiana glutinosa was obtained when the inoculum contained dianthin at 1–5 µg/ml.

Grasso et al. (1980) showed strong inhibition of tobacco mosaic virus multiplication in protoplasts by 10 µg of Phytolacca inhibitor/ml culture medium. Inhibition was complete within one hour of infection. The inhibitor also damaged the protoplasts, but this could be prevented by $CaCl_2$ treatment. Gessner and Irvine (1980) showed that the pokeweed inhibitor prevents the elongation of nascent polypeptide chains, by attacking a site on the 60S ribosomal sub-unit. This prevents translocation of aminoacylated tRNA from the acceptor site to the donor site. It is not surprising that protein synthesis in vitro by pokeweed ribosomes was not inhibited by the pokeweed inhibitors, whereas wheat, cowpea, and mammalian ribosomes were (Owens et al., 1973; Foa-Tomasi et al., 1982).

It seems likely that the antiviral activities of the dianthins and pokeweed inhibitors are expressions of their effects on protein synthesis (Irvin and Aron, 1982), although a separate mode of action cannot yet be excluded. Furthermore, the inhibitors do not protect the donor plants from viruses, as both Dianthus and Phytolacca can be infected. This raises an interesting question of evolution: why should one plant contain an inhibitor of virus infection of other species? Any ideas of evolutionary altruism should probably be dismissed out of hand, as in nature the inhibitor would never be present in the second species. One possibility is that the proteins have other functions in the donor species, and their effects on protein synthesis and virus infection in test species are secondary. However, the remarkable potency of the compounds is intriguing. Stirpe et al. (1981) have drawn some interesting biochemical and biological parallels with other plant proteins with cytotoxic or inhibitory properties, such as ricin and lectins. Clearly these groups of proteins have further interesting aspects to reveal, although their direct involvement in resistance to viruses is much in doubt.

Antiviral factors operating in genetically resistant hosts. The possible involvement of an antiviral factor (AVF) in the hypersensitive resistance to TMV of tobacco varieties containing the N-gene has been intensively studied. Sela et al. (1964) and Antignus et al. (1977) reported a factor from Nicotiana glutinosa which inhibited accumulation of TMV infectivity in N. tabacum cultivars. Reichman et al. (1983) showed by ELISA that AVF at 70 ng protein/ml caused a 60% inhibition of accumulation of TMV antigen. Suggestions that antiviral activity could be detected down to 0.1 ng/ml (Sela, 1981a) are not yet confirmed.

AVF has been partially purified and characterized as a protein. Several sugars were found associated with AVF, although some if not all were also found in 'mock' AVF preparations from healthy leaves (Mozes et al., 1978). The suggestion that AVF is a glycoprotein led

Fig. 6.6. Some possible ways in which the putative tobacco
 antiviral factor (AVF) might be involved in resistance to tobacco
 mosaic virus, and a scheme for AVF synthesis. Based on data in Sela
 (1981a; 1981b), and redrawn from Fraser (1985d).

to attempts to purify it by chromatography on
concanavalin-A-Sepharose, with elution by methyl glycosides. Antignus
et al. (1975) showed that partially purified AVF from infected leaves
appeared to be phosphorylated. Mozes et al. (1978) estimated the
molecular mass of the dephosphorylated protein to be 22 kDa.

After fractionation by various column chromatographic or gel
electrophoretic methods, and assay by chemical, radiochemical or
biological activities, the profiles of AVF reported were generally
very broad or heterogeneous, suggesting the presence of several types
of molecule. Sela (1981a) raised the possibility that this might
reflect different levels of phosphorylation or carbohydrate
composition, but other, contaminating proteins cannot be ruled out.

Sela and co-workers have proposed a complex scheme for the
synthesis, metabolism and possible antiviral activity of AVF, which
is summarized in Fig. 6.6. Some of the suggested steps are based on
experiments in which AVF-like activity was synthesized in vitro using

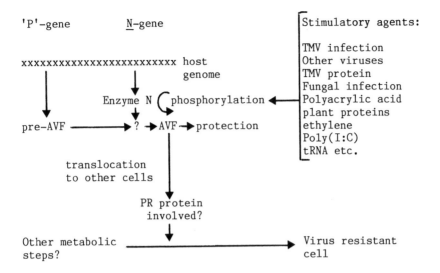

extracts from tobacco leaves of various genotypes. The scheme
proposes that a gene P, which is present in both N and n-gene leaves,
codes for a precursor to AVF (pre-AVF). The N-gene produces an enzyme
which converts pre-AVF to AVF (Sela and Harpaz, 1977; Sela et al.,
1978). It was suggested that phosphorylation is involved in the
activation of this enzyme; phosphorylation of AVF apparently did not
occur in vitro. However, it was suggested that phosphorylation of
AVF, and presumed activation, was promoted in vivo by TMV. This was
assumed to involve the double-stranded replicative forms of TMV,
because it could also be stimulated by synthetic, double-stranded
RNAs (Gat-Edelbaum et al., 1983), as well as by a number of other
biological and chemical agents.

Sela et al. (1978) and Gat-Edelbaum et al. (1983) suggested from
further in vitro experiments that there is a requirement for cyclic
AMP (cAMP) and GMP for activation. The possible roles of cyclic
nucleotides as regulatory molecules in plants have been the subject
of much argument, but there is now good evidence for their occurrence
(Brown and Newton, 1981). A problem is that assay methods are
required which will distinguish between the biologically active
3'-5'cAMP and the more common 2'-3'cAMP which is a
naturally-occurring breakdown product. The evidence presented by Sela
et al. (1978) showing that cAMP concentration was elevated in
TMV-infected plants was obtained by an indirect assay method.

It will be clear from this description of AVF that many features of
its metabolism and biological activity are conjectural. Some of its
described effects may be non-specific or admit of trivial
explanations. Furthermore, much of the published data for biological
activity shows comparatively low inhibitions of infectivity.

Throughout the development of research on AVF, parallels have been
drawn with the synthesis and mode of action of interferon in
mammalian cells. Thus there are similarities in the purification
procedures for AVF and interferon, and the role of double stranded
RNAs in induction of interferon is well known (De Maeyer et al.,
1971). The similarity became even closer following a report by
Orchansky et al. (1982) that mammalian interferon inhibited TMV

multiplication in tobacco leaf disks. This result was surprising, in view of the high species-specificity of mammalian interferons in animal cells. It was also unexpected that interferon appeared to show antiviral activity in plants at much lower concentrations than those required in animal cells. It is perhaps disturbing that in experiments with tobacco leaf disks, Reichman et al. (1983) found that human γ3 interferon was apparently more than 10,000 times as active as tobacco leaf AVF against TMV, on a molecule for molecule basis. This must suggest either that AVF is a comparatively inefficient inhibitor of virus multiplication in its own species, or that the AVF preparations tested were grossly contaminated or inactivated. Other workers have failed to detect inhibitory effects of mammalian interferons on plant viruses (Antoniw et al., 1984; Huisman et al., 1985; Loesch-Fries et al., 1985). Pierpoint (1983b) has also suggested that the apparent inhibition of TMV multiplication by human interferon in Orchansky's experiments might have been due to non-specific contaminants.

Fig. 6.6 indicates possible ways for expression of the antiviral effects of AVF. Again, some of these lean heavily on the interferon model, and the evidence for most is at best incomplete.

In mammalian cells, interferon induces synthesis of an oligonucleotide, $ppp(A2'p5')_n A$ (n = 2-4), usually abbreviated to 2,5-A. This activates a latent RNase which interferes with protein synthesis (Williams and Kerr, 1978). Devash et al. (1982) showed that synthetic 2,5-A inhibited TMV multiplication in tobacco leaf disks, while Reichman et al. (1983) suggested that plants contained an enzyme system capable of polymerizing ATP to 2,5-A in the presence of a double-stranded RNA inducer. These results suggested a possible mode of action of AVF (Sela, 1981b). But Cayley et al. (1982) could not detect 2,5-A or interferon-associated enzymes in higher plants, so the proposed role of 2,5-A in AVF activity is still in doubt.

Devash et al. (1981) have suggested another possibly antiviral effect which also involves a 2,5-A-like oligonucleotide. Crude preparations of AVF were reported to contain a 'discharging factor' (DF), which removed the histidine from the 3' (t-RNA like) end of

TMV, with possible effects on infectivity. Inactive DF preparations could have their activity restored by incubation with a 'polymerized ATP' fraction which the authors assumed to be like 2,5-A. Unfortunately, the possibility that this activity of DF might merely reflect non-specific nuclease activity was not fully examined. Furthermore, crude AVF preparations were shown to contain more nuclease activity than similar preparations from healthy plants.

There have been two further suggestions which both involve indirect action of AVF. Because of the allegedly low concentrations required for effect (Mozes et al., 1978), it has been suggested that AVF might act as a hormone rather than a direct antiviral agent. Nothing appears to be known about how AVF might induce the postulated antiviral state in treated cells. Sela (1981b) has also suggested that AVF might operate through an effect on pathogenesis-related proteins. Again the mechanism is purely conjectural, and no antiviral activity has been proved for the PR proteins.

If AVF is important in the control of the localization of TMV in tobacco, it should be possible to demonstrate that it is synthesized in sufficient quantity and at the right time. Little attention appears to have been given to this aspect. AVF was detectable by labelling with radioactive phosphorus by 48 h after inoculation, but little was found at 24 h (Antignus et al., 1977). This contrasts with other results suggesting that inhibition of TMV multiplication is already well established by 24 h (Otsuki et al., 1972). This might suggest that AVF is unlikely to be a critical or early event in TMV localization.

Loebenstein and Gera (1981) reported a different type of antiviral agent from N-gene tobacco, which they named 'inhibitor of virus replication' (IVR). The origin and properties suggest no immediate similarities with AVF, although parallels have again been drawn with interferon. IVR consisted of two proteins of molecular masses 26 kDa and 57 kDa. They were found in the culture medium of TMV-infected protoplasts from N-gene plants, but were not produced by healthy protoplasts, or by TMV-infected protoplasts from n-gene plants. Assay of antiviral activity was by using infected protoplasts, and following changes in infectivity or viral antigen. Even the most concentrated preparations of IVR gave only 60% inhibition, and possible non-specific inhibitory effects on metabolism were not investigated. Indirect supporting evidence for an antiviral role was obtained by Gera et al. (1983), who showed that treatment of protoplasts from N-gene plants with actinomycin-D or chloramphenicol increased TMV multiplication but reduced accumulation of IVR.

6.4.5. Localization mechanisms: conclusions

It is clear that localization of a virus infection, with or without the formation of a necrotic lesion, involves complex changes in many biochemical processes. In the present state of knowledge, it is not possible to say which are directly involved in localization, and which are secondary. A possible way to distinguish these two classes of response is to examine the relative timing of events. Fig. 6.4 summarizes some of the major changes, and relates them to known markers of infection, localization and necrosis. It has to be stressed that the Figure combines results from different laboratories, operating under different conditions with various combinations of hosts and viruses; thus it provides only a general guide to comparative timings.

Some of the major changes, such as cell death or wall thickenings, occur too late to be primary events in localization. There is evidence that some of the changes in plant growth regulator metabolism are also secondary, although the well-characterized changes in ethylene metabolism seem to be intimately associated with the early onset of necrogenesis. What is interesting is that there appears to be a 'cluster' of phenomena which are early and may be associated with membrane changes: these include increased electrolyte leakage, early loss of the requirement for cell-to-cell contact, increased ethylene biosynthesis and some effects of inhibitors and membrane stabilizers. However, there is still no explanation of how these might be involved in localization.

The various compounds with demonstrable antiviral activity suffer from incomplete understanding of their modes of action, and lack of proof that some of the effects are specifically antiviral. There is generally a need for further control experiments, and some of the parallels with interferon have perhaps been given too much emphasis.

6.4.6. Resistance mechanisms permitting some spread of virus

Some resistance mechanisms do not operate by localization around the site of infection, but permit at least some spread of the virus, either within the inoculated leaf or systemically. These mechanisms might inhibit virus multiplication, or rate of long-distance transport, or development of visible symptoms, or more than one of these.

The term 'tolerance' is frequently used to describe examples where visible symptoms of infection are reduced or absent, while virus multiplication is detectable. The implication is that such resistance takes the form of insensitivity to the pathogenic effects of infection, without direct inhibition of multiplication. However, in some cases virus multiplication has been detected but not quantified, and the effects of resistance on multiplication have not been excluded. 'Tolerant' plants with no visible mosaic symptoms may also show reduced yield compared with healthy plants (Kooistra, 1968).

With localizing resistance, the target of the resistance mechanism is to some extent specified by the phenotype: it must prevent multiplication or cell-to-cell spread of the virus within a short distance from the point of infection. Non-localizing resistance mechanisms have a wider range of potential targets. These could include partial inhibition of any of the stages of virus multiplication, incomplete activity against cell-to-cell spread, or some interference with longer-distance movement of the virus, for example by preventing access to the vascular system. Some of these kinds of activity can be excluded for particular host and virus combinations by testing whether resistance operates in protoplasts. However, our understanding of the target of non-localizing resistance mechanisms remains comparatively diffuse; this may explain why these mechanisms have been the subject of less research than those involving localization, as it is harder to erect testable hypotheses.

Resistance to CMV in cucumber. This is one of the more intensively studied systems, although the genetic basis of resistance is not very clear. Cultivars have been used with resistance derived from either of two sources: Chinese Long (Wasuwat and Walker, 1961; Amemiya and Misawa, 1977), or Kyoto 3-foot (Levy *et al*., 1974; Maule *et al*., 1980). The mechanisms may be genetically distinct (Coutts and Wood, 1977), but no major differences in mode of action have yet been found.

In whole plants, resistance suppressed the development of mosaic symptoms and allowed only one-tenth as much virus multiplication as in susceptible plants (Wood and Barbara, 1971; Amemiya and Misawa, 1977). No differences in the pattern of virus accumulation could be detected until 36 h after inoculation. Furthermore, development of resistance was inhibited by treatment with actinomycin-D (Nachman *et al*., 1971; Barbara and Wood, 1974), α-amanitin (Amemiya and Misawa, 1977) or ultraviolet light (Levy *et al*., 1974). All three inhibitors increased multiplication of CMV in resistant plants only if applied not more than 12 to 24 h after inoculation, and none inhibited virus multiplication in susceptible plants. These results suggest that early synthesis or activation of a virus inhibitor, dependent on transcription of DNA, is involved in the development of resistance.

Protoplasts from resistant (Kyoto) plants produced less progeny virus when inoculated, and a smaller proportion of detectably infected cells, than protoplasts from a susceptible cultivar (Maule *et al*., 1980). Protoplasts from resistant (China) plants also appeared to show resistance. These results suggest that the resistance mechanism operates within the individual cell, and does not act against cell-to-cell spread of the virus. Resistance was also shown to operate when protoplasts were inoculated with CMV RNA. This suggests that attachment or uncoating of virus particles is not the target.

An interesting difference between whole-leaf and protoplast studies was that protoplasts did not allow an early phase of unrestricted virus multiplication (Maule *et al*., 1980). Furthermore, treatment with actinomycin-D or ultraviolet light did not reduce the inhibition of virus multiplication in protoplasts (Boulton et al., 1985). This led to the suggestion that there might be two components of resistance in leaves: a constitutive one at the level of the individual cell, which limits virus multiplication, and one induced after infection and requiring some intercellular level of response. Only the first is thought to operate in isolated protoplasts.

Various differences in metabolism between resistant and susceptible plants have been noted after infection, including changes in respiration (Menke and Walker, 1963), in peroxidase isoenzymes and in polyphenoloxidase (Wood, 1971; Wood and Barbara, 1971). None of these differences appeared to correlate with resistance. Cucumber plants have also been shown to contain an inhibitor which reduces the numbers of local lesions formed by CMV on cowpea (Sill and Walker, 1952), but resistance in cucumber was not related to differences in inhibitor concentration (Wasuwat and Walker, 1961).

Resistance in Phaseolus vulgaris to BCMV. Genetically, this is amongst the most complex systems of resistance yet investigated. Apart from a virus-localizing mechanism based on the dominant I gene, resistance is controlled by a recessive gene system involving several loci (Drijfhout, 1978). This system is also one of the few where there is good evidence that resistance requires cooperative action between genes at different loci. Genes at the bc-1, bc-2 or bc-3 loci (collectively termed the bc-x loci) confer resistance against specific virus strains. As judged by symptom suppression, bc-x genes are only effective when in combination with the bc-u gene, which is non-specific for virus strain. Some of the genetic interactions are summarized, in abbreviated form, in Table. 6.2. This also illustrates how virus isolates with particular virulence genes, or combinations of virulence genes, can overcome specific resistance genes.

Day (1984) examined the effects of different host genotypes on virus multiplication in more detail, in an attempt to define targets and functions for each type of gene. Plants homozygous for bc-u, but containing susceptible alleles at all the bc-x loci, had no detectable resistance, confirming that bc-u had no antiviral activity

Table 6.2. Genetic interactions between <u>Phaseolus</u> <u>vulgaris</u> plants with different susceptibility or resistance genes at the <u>bc-u</u> and <u>bc-1</u> loci, and strains of bean common mosaic virus with different virulence genes.

Host genotypes[a]	Virus genotypes			
	observed			theoretical
	0	1	1.1^2	1^2
$bc\text{-}u^+, bc\text{-}u^+/bc\text{-}1^+, bc\text{-}1^+$	M	M	M	m
$bc\text{-}u^+, bc\text{-}u^+/bc\text{-}1, bc\text{-}1$	M/R	M	M	m/r?
$bc\text{-}u^+\, bc\text{-}u/bc\text{-}1, bc\text{-}1$	M	M	M	m
$bc\text{-}u, bc\text{-}u/bc\text{-}1^+, bc\text{-}1$	M	M	M	m
$bc\text{-}u, bc\text{-}u/bc\text{-}1^+, bc\text{-}1^+$	M	M	M	m
$bc\text{-}u, bc\text{-}u/bc\text{-}1, bc\text{-}1$	R	M	M	r
$bc\text{-}u, bc\text{-}u/bc\text{-}1^2\, bc\text{-}1^2$	R	R	M	m
$bc\text{-}u, bc\text{-}u/bc\text{-}1^+\, bc\text{-}1^2$	M	M	M	m
$bc\text{-}u, bc\text{-}u/bc\text{-}1, bc\text{-}1^2$	R	R	M	r

[a]Not all possible heterozygotes are shown. R,r = resistant, M,m = mosaic. Based on the results of Drijfhout (1978) and Day (1984).

alone. Plants with the genotype $bc\text{-}u^+, bc\text{-}u^+/bc\text{-}1, bc\text{-}1$ showed detectable systemic infectivity but no symptoms. This suggests that the <u>bc-1</u> allele alone might prevent symptom formation. Plants heterozygous at both the <u>bc-u</u> and <u>bc-1</u> loci did not suppress symptoms, but contained less virus than homozygous susceptible plants. Thus the genes are fully recessive for symptom suppression, but show gene-dosage effects on virus multiplication.

Plants with homozygous <u>bc-u</u> plus <u>bc-1</u> or <u>bc-2</u> did not appear to suppress virus multiplication in the inoculated leaves. This might suggest either that a comparatively long period is required for

140

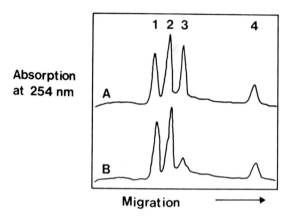

Fig. 6.7. Polyacrylamide gel electrophoresis of CCMV RNAs purified from virus multiplied in (A) susceptible and (B) resistant cowpeas. Reproduced from Wyatt and Kuhn (1979) by permission of the authors and the American Phytopathological Society.

induction of resistance, or that resistance does not operate against cell-to-cell spread of the virus. No virus could be detected in the petioles of inoculated resistant leaves, suggesting that the resistance may have prevented loading of the virus into the phloem, in which BCMV is translocated (Ekpo and Saettler, 1975). Bijaisoradat and Kuhn (1985) also found a high level of multiplication of virus in inoculated leaves, and very low systemic spread, in certain resistant lines of soybean infected with CCMV.

Resistance to CCMV in cowpea. This is a third example of resistance which apparently operates by preventing systemic spread of virus from the inoculated leaf (Wyatt and Kuhn, 1979), but in this case the rate of virus multiplication was found to be significantly lower in inoculated resistant leaves than in susceptible leaves. This implies inhibition of both virus multiplication and movement. Virus multiplied normally in protoplasts isolated from resistant leaves (Wyatt and Wilkinson, 1984). Thus the resistance mechanism operating in whole plants was either lost during the preparation of protoplasts, or it depends on cooperation between cells.

The virus produced in resistant plants had comparatively small amounts of the genomic RNA-3 (Fig. 6.7), a deficiency which disappeared on transfer to a susceptible host. CCMV RNA-3 contains two genes, one of which codes for coat protein. However, this is not expressed directly in vivo (Lane, 1981): coat protein is translated from a monocistronic messenger, RNA-4. CCMV from resistant plants contained normal amounts of RNA-4, so this sequence and synthesis of viral coat protein are unlikely to have been targets of the resistance mechanism. The other gene on RNA-3 codes for a protein of molecular mass 35 kDa, which by analogy with the related virus BMV might be involved in the replicase function (Hariharasubramanian et al., 1973). This might explain an effect of resistance on virus multiplication. However, complementation experiments with resistance-breaking isolates (considered in section 6.5) suggest that a function specified by RNA-1 might be separately involved in resistance to virus movement (Wyatt and Kuhn, 1980).

Resistance to TMV in tomato. Interpretation of experiments on resistance in the three systems mentioned above is made somewhat difficult by the fact that quite different resistant and susceptible cultivars have to be compared. Any difference, especially a quantitative one, might be a result of host genetic background rather than of direct expression of the resistance gene. In contrast, TMV resistance in tomato offers an opportunity to minimize the possible effects of host genetic background, and to isolate metabolism associated with a particular resistance gene. Firstly, Darby et al. (1978) have bred nearly isogenic lines containing different combinations of the Tm-1, Tm-2 and Tm-2^2 resistance genes. Secondly, each gene is also available in different host genetic backgrounds. The genetics of the interaction between the host resistance genes and corresponding genes for virulence in the virus have been discussed by Pelham (1972) and Hall (1980).

Most biochemical studies have concentrated on Tm-1. In plants grown at different temperatures, inhibition of virus multiplication was

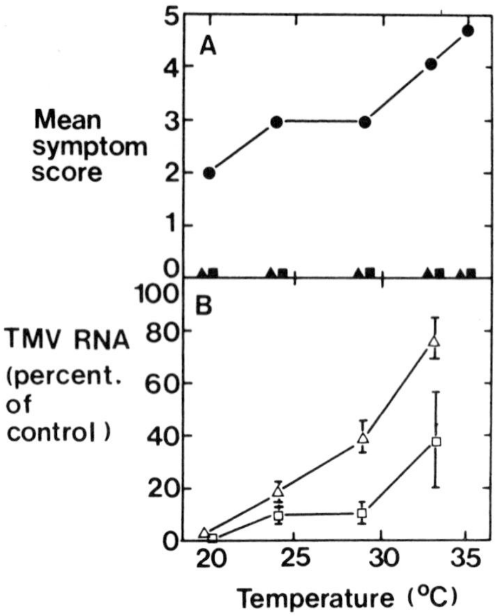

Fig. 6.8. Effects of the _Tm-1_ gene on mosaic symptom formation and multiplication of tobacco mosaic virus, in tomato plants grown at different temperatures. (A). Symptom severity (assessed on a scale of 1 - 5) in susceptible plants (●), and plants heterozygous (▲) or homozygous (■) for the the _Tm-1_ resistance gene. (B). Virus multiplication, measured as TMV RNA accumulation, and expressed as a percentage of the level in susceptible plants grown at the same temperature, for plants heterozygous (△) or homozygous (□) for the _Tm-1_ gene. From Fraser and Loughlin, 1982.

always more effective in plants with homozygous _Tm-1_ than in those with the gene in the heterozygous configuration (Fraser and Loughlin, 1982). Inhibition decreased with increasing temperature of growth (Fig. 6.8). Thus the effects of _Tm-1_ on TMV multiplication are gene-dosage dependent and temperature sensitive. In contrast, the gene was completely effective in suppression of mosaic symptom formation in the heterozygous configuration, at all temperatures tested (Fig. 6.8). Thus the symptom-suppression function of the gene shows straight dominance and is temperature-insensitive. One possible explanation is that the 'gene' actually consists of two independent but closely-linked genetic functions. Another is that the pathway between gene and phenotypic effects might be branched, with separate functional end-products affecting multiplication and symptom suppression. However, it is possible that a single functional product could give the apparently contradictory results observed, if the kinetics of interaction leading to the two phenotypic effects are different. Kacser and Burns (1981) provide a useful mathematical and biochemical model for such genetic effects.

In plants containing the Tm-1 gene, there was a delay of 8 to 40 days between inoculation and the first detection of increased infectivity; the delay was greater in those plants with homozygous Tm-1 (Dawson, 1965; Fraser and Loughlin, 1980). This suggests that the inhibitor of virus multiplication is present in uninfected plants and does not have to be induced by an early phase of virus multiplication. This conclusion is supported by the observation that TMV multiplication was inhibited in protoplasts from Tm-1 plants (Motoyoshi and Oshima, 1977). Their results also showed that the resistance mechanism operates within individual cells, and does not act by preventing cell-to-cell spread of virus. Tm-1 resistance was effective when protoplasts were inoculated with TMV RNA (Motoyoshi and Oshima, 1979), suggesting that resistance must operate at some stage after virus uptake and uncoating.

Evans et al. (1984; 1985) measured the levels of RNA-dependent RNA polymerases in susceptible and resistant plants. Healthy plants of both genotypes contained high levels of polymerase activity, but these were not able to synthesize the replicative form and replicative intermediates of TMV in vitro when supplied with exogenous TMV RNA as template. Infection with TMV strain 0, which does not overcome Tm-1 resistance, caused a two- to four-fold increase in total RNA-dependent RNA polymerase activity in susceptible, but not in Tm-1 plants, and this increased activity was able to radiolabel the TMV RNA replicative form and replicative intermediate in vitro. These results could suggest that the Tm-1 gene product operates by preventing synthesis or modification of the enzyme activity required to form a functioning TMV replicase, but they do not exclude more trivial explanations.

Mature plant resistance. Some cases are known where plants of normally susceptible cultivars become more resistant to viruses during maturation. In potato, Venekamp and Beemster (1980) and Venekamp et al. (1980) showed that resistance to systemic infections of PVY and PVX increased as plants aged. They correlated increased resistance with reduction in leaf ribosome concentration. Resistance in this example was likely to have been non-specific, reflecting a reduction in the ability of aged leaves to support normal amounts of virus multiplication. Fraser (1981) also noted that tobacco leaves became much more resistant to infection by TMV during flowering and early senescence. The change was ascribed to increasing toughness and resistance to mechanical wounding.

6.5. THE BIOCHEMISTRY OF VIRULENCE

6.5.1. Theoretical models and practical approaches
A virulence gene is expressed as the ability of a particular virus isolate to multiply or cause symptoms in a plant containing a gene for resistance to that virus. Table 6.1 gives data on the frequency of virulence against different types of resistance gene. Acquisition of virulence could involve two types of change in the virus (see Fig. 6.1).

A qualitative change in some viral component could result in the failure of an interaction with a host component. This interaction might occur at the stage of recognition and induction of a host resistance mechanism. It might also be by failure to interact with a pre-formed or induced antiviral compound. Alternatively, a qualitative change in a virus component might lead to successful interaction with some host susceptibility function, with which the parent strain of virus does not interact. This model of virulence overlaps with possible mechanisms determining pathogenicity and host range.

Virulence might also involve a quantitative change in virus

multiplication or activity of a viral function. This could lead to faster virus spread or multiplication, against which the host defence mechanism is too slowly induced. Greater virus spread or multiplication might also swamp limited amounts of host antiviral factors.

In contrast to plant functions such as resistance, where the difficulty is in disentangling a single biochemical aspect from the background of tens of thousands of other reactions, virulence in the virus should be comparatively easy to study. The virus, in its various forms, can be completely characterized at the molecular level by the base sequences of its nucleic acids and amino acid sequences of the possible derived proteins. However, mere comparison of nucleic acid base sequences for avirulent and virulent isolates cannot give a complete explanation of virulence; it is still necessary to relate virulence to specific functions, and to explain the basis of the altered interaction with (unknown) host components. Furthermore, the high error frequency of copying of RNA genomes (Reanney, 1982) suggests that there could be difficulties in distinguishing sequence differences related to virulence, from those caused by random noise, if sequences of cDNA clones are compared. The available approaches are either to sequence a number of clones to obtain a consensus sequence, or to extract this directly by RNA sequencing.

Site-directed mutagenesis of cDNA clones should ultimately allow probing of the fine structure of the genetics of virulence. To date, however, most workers have attempted only a coarser type of genetic analysis, or have tried to characterize virulence at the level of altered function of gene products.

6.5.2. Pseudorecombinants and the genetic analysis of virulence

The individual genomic components of viruses with split genomes can generally be separated by centrifugation or electrophoresis. Complete, infectious genomes can then be reassembled using components from virus strains differing in virulence. These 'pseudorecombinants' allow mapping of virulence to particular genomic segments.

Table 6.3 shows several examples where determinants of virulence

Table 6.3. Location of genes for virulence or for host range control in viruses with multicomponent genomes.

Virus strains[a]	Location of virulence or pathogenicity genes	Non-host or resistant line infected
RRV-LG RRV-S (nepovirus)	RNA-1 of LG	Raspberry 'Lloyd-George'
CLRV-R CLRV-G (nepovirus)	RNA-1 of R	Gomphrena globosa
PNRV-S PNRV-G (ilarvirus)	B particles of S	Vigna sinensis
CCMV-R CCMV-T (bromovirus)	RNA-1 of R controls systemic invasion; RNA-1 and RNA-3 of R control replication	Vigna sinensis
CMV-B CMV-LsS (cucumovirus)	RNA-2 of B	various legumes
CMV-LsS CMV-B	RNA-2 and RNA-3 of LsS	Lactuca saligna

[a]The first of each pair of virus strains will infect the plant; the second will not, or the plant is resistant to it. Pseudorecombinants containing the specified components of the first (hosted) strain, plus the remaining components of the other (non-hosted) strain, will infect the plant or overcome the resistance. References: RRV, Harrison et al. (1974); CLRV, Jones & Duncan (1980); PNRV, Loesch & Fulton (1975); CCMV, Wyatt & Kuhn (1980); CMV, Edwards et al. (1983).

have been mapped. It also includes examples from an overlapping problem: the determinants of host range. The viruses involved come from diverse groups, and it cannot be assumed that there is necessarily any similarity in function between genomic segments, or in distribution of functions between segments. Thus the fact that virulence (or host range) determinants appear on different RNAs in different virus groups does not mean that virulence is necessarily associated with different functions. However, virulence or host range control can involve more than one genomic segment of an individual virus. Thus in CMV, two genetic determinants are required for systemic invasion of Lactuca saligna, whereas only one segment determines the ability to multiply in various legumes (Edwards et al., 1983).

Association of genetically-mapped virulence with altered gene products is at present hampered by ignorance of the products of most viral genes, or of their functions. It does appear from the limited number of studies with pseudorecombinants, that virulence is more often associated with genomic segments which do not contain the coat protein gene.

6.5.3. Other approaches to the biochemistry of virulence

Genetic mapping of viruses with monopartite genomes is much more difficult, but some oblique approaches have been used. Kado and Knight (1966) studied the interaction of TMV with the N' resistance gene in Nicotiana sylvestris. They chose a virus strain which spreads systemically, and stripped the coat protein from the 5' end to varying degrees by treatment with sodium dodecyl sulphate. RNA thus exposed is more sensitive to mutation by nitrous acid than RNA which is still covered by coat protein. The numbers of mutants to the local lesion phenotype (i.e. to avirulence) rose when about 70% of the RNA had been exposed (Fig. 6.9). This region of the genome codes for a 30 kDa protein, which may be involved in cell-to-cell spread of the virus (Nishiguchi et al., 1978; Ohno et al., 1983). Kado and Knight's results would tend to exclude the coat protein gene as the determinant of local lesions, as it is nearer to the 3' end, However,

148

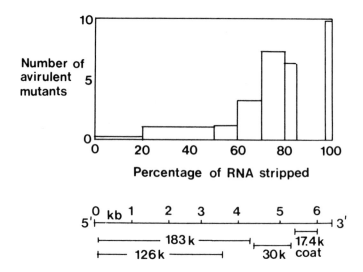

Fig. 6.9. Mapping of the determinant of the local lesion reaction on N'-gene *Nicotiana* plants on TMV RNA. Coat protein subunits were progressively stripped from the 5' end of particles causing a systemic infection, and the frequency of local-lesion mutants determined after mutagenesis with nitrous acid. Based on data from Kado and Knight (1966). The lower scale shows the genetic map of TMV RNA with the known principal *in vivo* translational products.

one must question whether the rather indirect method used for mapping would allow this level of discrimination.

Kado and Knight also noted evidence from other sources which might implicate the coat protein as a determinant. Tsugita and Fraenkel-Conrat (1960) and Tsugita (1962) mutagenized a systemic strain of TMV, and selected local lesion mutants on N'-gene tobacco. Almost all of their local lesion mutants had coat proteins with altered amino acid compositions. Furthermore, only one of 60 different mutants, which still gave systemic symptoms on N'-gene plants, differed from the parent wild type in amino acid composition of the coat protein.

Separate evidence suggesting an involvement of coat protein in the interaction with the N'-gene comes from a study of the sizes of lesions caused by different isolates of TMV (Fraser, 1983). There was no correlation between lesion size and the intrinsic ability of each isolate to multiply, as measured in a systemic host. There was, however, a strong correlation between lesion size and the temperature-sensitivity of the coat protein subunits. Temperature sensitivity was chosen simply as an indicator of biochemical differences, and does not imply that thermal denaturation of coat protein subunits plays any part in determination of the hypersensitive reaction. The results, together with those of Tsugita (1962), suggest that coat protein might at least modify the operation of the local lesion reaction. Furthermore, by extrapolation from the results with pseudorecombinants, there is the possibility that virulence in viruses with monopartite genomes also might be controlled at more than one locus.

Coat protein properties have also been examined in connection with the ability of TMV to overcome the Tm-1 resistance gene in tomato. Dawson et al. (1975) reported that the amino acid compositions of two resistance-breaking strains differed from the avirulent wild types. Later Dawson et al. (1979) examined a larger number of strains and suggested that there was no correlation between the ability to overcome Tm-1 and the amino acid composition. However, this study was directed mainly at the detection of methionine which is not usually found in TMV coat protein. The data as presented suggest that all resistance-breaking isolates probably did differ in amino acid composition in other ways. Fraser et al. (1983) also found that twenty strain 1 isolates differed from strain 0 in coat protein properties. Many of these isolates had been prepared from the strain 0 parent by nitrous acid mutagenesis under conditions designed to give a low mutation frequency. This would tend to minimize the possibility of coat protein mutations occurring by chance, and in addition to mutation in a virulence determinant elsewhere in the genome.

McRitchie and Alexander (1963) found that TMV isolates capable of breaking Tm-1 resistance could be differentiated serologically from strain 0 isolates. Valverde and Fulton (1982) also found that isolates of SBMV capable of overcoming the localizing resistance in

Fig. 6.10. Polyacrylamide gel electrophoresis of AMV particles from cowpea protoplasts inoculated with various combinations of AMV component particles. The wild-type virus caused local lesions on cowpea plants; the mutant spread systemically. (A). Inoculated with mutant particles at 3 μg/ml. (B). Wild-type particles at 3 μg/ml. (C). Mutant particles at 2 μg/ml plus wild-type M component at 1 μg/ml. (D). Wild type particles at 2 μg/ml plus wild type M-component at 1 μg/ml. Ta, Tb, M and B indicate Top a, Top b, Middle and Bottom components. From Roosien et al., 1983, by permission of the authors and Martinus Nijhoff/Dr. W. Junk Publishers.

Phaseolus vulgaris could be distinguished serologically from the avirulent type strain. It is not clear whether these differences reflect a role of coat protein in virulence, or merely random variation.

Another use of mutation to probe virulence was developed with AMV in cowpea (Roosien et al., 1983). The wild-type parent caused a local lesion reaction, whereas the mutant spread systemically, and was also replicated to a higher level in protoplasts. AMV has a divided genome packaged in various particles (Ta, Tb, M and B) which can be separated by centrifugation. The mutant virus synthesized in protoplasts contained a higher proportion of the middle (M)

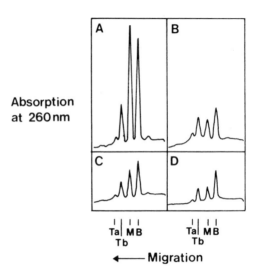

component, which contains RNA-2, than the wild-type parent (Fig. 6.10). Furthermore, addition of wild-type middle component to the mutant caused wild-type infection patterns in plants and protoplasts. It can be argued that wild-type RNA-2 inhibited virus multiplication in protoplasts and activated the local lesion response on leaves.

Two studies with artificial mutants or natural variants of CPMV have also cast some light on the nature of virulence. A nitrous acid mutant prepared by De Jager and Van Kammen (1970) caused chlorotic local lesions on Phaseolus vulgaris at 22°C, unlike the parent wild-type which spread systemically. Thus the mutation involved was apparently one to avirulence. At 30°C, both parent strain and mutant spread systemically, so it was unlikely that the mutant was defective unless it was cold-sensitive. At 22°C, the mutant formed excess amounts of top component particles. Experiments with pseudorecombinants between the mutant and wild-type showed that the middle component determined virulence. It was suggested that the middle component might control the formation of excess top component and the inhibition of systemic spread.

Two natural isolates of CPMV studied by De Jager and Wesseling (1981) differed from normal in that they did not induce a local lesion reaction on cowpea cv. Early Red. Both spread systemically; one induced necrosis and the other mosaic. This shows that there are two separate determinants of virulence. These are expressed as the ability to evade localization, and the ability to induce or not to induce necrosis. Analysis of pseudorecombinants showed that the determinant of disease type is located in the bottom component.

A final approach to the study of virulence is by complementation analysis, where avirulent and virulent isolates are inoculated together to resistant plants. There seems to have been little attempt to use this to map or identify virulence functions. Taliansky et al. (1982) did show that in tomato plants with the resistance gene Tm-2, which appears to localize or severely restrict the spread of TMV, infection with PVX permitted systemic invasion by TMV. It was not clear whether this involved true complementation or disabling of a host TMV resistance mechanism by PVX. In either case, the biochemistry deserves further study.

6.6. INDUCED RESISTANCE

6.6.1. The diversity of induced resistance mechanisms

It is clear from the literature that there are many different types of mechanism of induced resistance. In some, the resistance can be induced by a wide range of biotic and abiotic factors, including viruses, fungi, bacteria, chemicals and host developmental factors. Some types of induced resistance are non-specific, in that the resistance induced is effective against a wide range of viruses, microbial pathogens and even insects (McIntyre et al., 1981). In others, the effect can be highly specific, with resistance expressed only against the same virus as the inducer, or indeed only against certain related isolates.

Induced resistance can be localized, in that it is restricted to the tissue surrounding the initial infection or chemical treatment, or it can be systemic. Different types of resistance can be induced by virus infections which are themselves localized, or by systemically-spreading infections. Plants with systemic infections which fail to invade certain regions, such as islands of dark green tissue, may have different types of induced resistance mechanism in the infected and non-infected tissue.

These complications stress the importance of examining the diversity of induced resistance mechanisms, and of not applying conclusions from one experimental system to another. In the following discussion, known examples of induced resistance are grouped in three main types, but it is stressed that individual host and virus combinations within each group need not all share the same properties.

6.6.2. Localized acquired resistance, acquired systemic resistance and the pathogenesis-related (PR) proteins

Viral infection often leads to localization of the virus in necrotic lesions; this is, in itself, a genetically-controlled resistance response (see section 6.4). It was recognized at an early stage that plants with lesions, when challenge-inoculated with the same or other lesion-forming viruses, either did not form further lesions in the vicinity of the primary necrotic response, or that those which did develop were smaller and less numerous. This is referred to as localized acquired resistance (Yarwood, 1960; Ross, 1961a). Most reports suggested that such resistance is non-specific (Hecht and Bateman, 1964), although there are exceptions (Yarwood, 1960).

In other reports, a systemic response to the initial lesion formation was also shown, in that the lesions formed after challenge inoculation of the upper leaves were smaller, or less numerous, than those formed on previously uninoculated plants (Gilpatrick and Weintraub, 1952; Ross, 1961b). This effect is referred to as acquired systemic resistance, and generally appears to be non-specific (Ross, 1966).

The biochemistry of localized acquired resistance has received little attention, whilst acquired systemic resistance has been studied intensively. Obviously, they might involve similar mechanisms, merely differing in degree of spread in the plant. Alternatively, localized acquired resistance might represent activation of the primary localization response, with acquired systemic resistance as a separate phenomenon. Loebenstein *et al.* (1968) showed that actinomycin-D inhibited the occurrence of localized acquired resistance to TMV in tobacco and bean, suggesting that transcription of DNA was required for induction of localized resistance. But the same inhibitor also had direct effects on the primary infection (Loebenstein *et al.*, 1969), so a specific effect on the induced, as opposed to the primary resistance,is difficult to justify.

A stimulus to the study of acquired systemic resistance,came with the discovery of the 'pathogenesis-related' (PR) or 'b' proteins, and suggestions that they might be involved in the induced resistance (Kassanis *et al.*, 1974; Gianinazzi, 1982). As with the antiviral factors considered in section 6.4.4, parallels were drawn between PR proteins and the antiviral effects of interferon in animal cells.

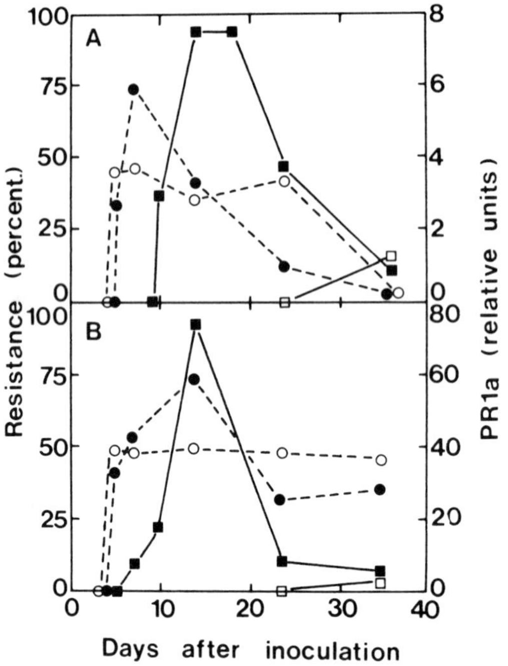

Fig. 6.11. Changes in the amounts of acquired systemic resistance and the concentrations of the PR1a pathogenesis-related protein in Xanthi-nc tobacco leaves. On day 0, plants were inoculated on one half of two lower leaves with TMV at a concentration sufficient to cause 50–100 lesions/half leaf. Control plants were sham-inoculated. The concentration of PR1a was measured at various times afterwards in upper leaves (A) of control (□) and inoculated (■) plants, and (B) in the previously untreated halves of the lower leaves of (□) control and (■) inoculated plants. Acquired systemic resistance was measured in (A) the upper leaves and (B) the previously untreated halves of lower leaves, and expressed as a percentage reduction in either lesion number (●) or diameter (○) caused by the primary inoculation. From Fraser (1982).

From studies with a number of combinations of hosts and pathogens, it became evident that PR proteins, and acquired systemic resistance, can be induced by numerous agents, and that the proteins and induced resistance tend to occur together. Some of the biotic and abiotic factors which induce both acquired systemic resistance and PR-like proteins are summarized in Table 4.1. Further examples are given by

Fraser (1985a) and Van Loon (1985).

The biochemical properties of PR proteins and their induction are covered in Chapter 4.5. Their possible involvement in the primary localization of virus infection is discussed in section 6.4.3. The objective here is to assess whether PR proteins are involved in acquired systemic resistance, to consider alternative functions for the proteins, and other explanations of the induced resistance.

If PR proteins are involved in acquired systemic resistance, there should be some sort of temporal and quantitative relationship between the proteins and resistance. Furthermore, neither should occur in the absence of the other. Fig. 6.11 shows changes with time in amounts of the commonest PR protein, PR1a, in Xanthi-nc tobacco, after inoculation of one half of each of two lower leaves. PR1a concentration was assessed in the opposite, uninoculated half of these leaves, and in upper, uninfected leaves. At the same time, further plants were challenge-inoculated with TMV to measure acquired resistance, expressed as reduction in lesion size or number. There was not a good correlation between amounts of PR1a and amounts of resistance, and the relative timings of changes in protein and resistance also pose problems for the hypothesis. For example, amounts of resistance induced in lower and upper leaves were similar, but the amounts of PR1a differed by an order of magnitude. Furthermore, acquired resistance could be detected before PR proteins.

In healthy Xanthi-nc tobacco plants which were beginning to flower, and on which the lower leaves had started to senesce, PR proteins accumulated without experimental treatment (Fraser, 1981). Removal of either or both of the developing inflorescence and the senescing lower leaves inhibited accumulation of PR1a, but no correlation was observed between the concentration of the PR protein and the sizes or numbers of lesions formed.

Both sets of data argue against any simple link between occurrence and amount of PR protein, and the amount of induced resistance. Fraser (1982) and Van Loon and Antoniw (1982) have reported experiments where effects similar to acquired systemic resistance could be induced in the absence of PR proteins. Fraser (1982) and Fraser et al. (1979) have described situations where PR-like proteins appeared to accumulate without the induction of systemic resistance. Finally, Kassanis and White (1978) were unable to demonstrate any reduction in TMV multiplication when purified PR proteins were added to infected protoplasts. All these lines of evidence suggest that PR proteins might not be directly involved in acquired systemic resistance. However, it is stressed that it is logically impossible to prove such a negative statement: failure to demonstrate an involvement might always have been a result of inadequate experimental conditions or of testing an inadequate hypothesis of the mechanism of involvement.

What other mechanisms might be involved in the control of reduced lesion size and number in acquired systemic resistance? The first point to make is that the two effects seem to be under controls which are at least partly separate. Thus the time-course of development of acquired resistance expressed as reduction in lesion size was quite different from that for reduction in lesion number (Fig. 6.11). Furthermore, Nicotiana glutinosa plants with acquired systemic resistance, measured as reduction in lesion diameter, actually formed many more lesions than control plants (Fraser et al., 1979). Ross (1966) and Balazs et al. (1976) suggested that reduction in lesion number might be a result of reduced size, with lesions becoming too small to see and count; this argument does not hold in all cases and does not disprove separate controls.

Alteration in lesion numbers as a result of the primary inoculation might be due to a change in susceptibility to mechanical wounding. However, there do not appear to have been any investigations of changes in the thickness of the cuticle and epidermal wall.

There is some evidence that altered lesion numbers may be related to changes in leaf water status as a consequence of the treatment which was used to induce resistance. Turgid leaves are generally more susceptible to wounding during mechanical inoculation than flaccid ones (Yarwood, 1959). Cassells et al. (1978) treated plants with an antitranspirant, and showed that this prevented the reduction in lesion number caused by a previous treatment with polyacrylic acid. This experiment has been criticized on the grounds that the concentration of polyacrylic acid used to induce resistance was very high (Kassanis, 1981), but this criticism does not invalidate the conclusion. Other work has shown changes in abscisic acid (ABA) concentration in association with either decreases or increases in susceptibility to infection after a primary inoculation. ABA concentration is an indicator of leaf water stress, and the correlations between the implied alteration in water status and changed lesion numbers are good (Fraser _et al_., 1979; Whenham and Fraser, 1981). These experiments also showed considerable transport of ABA from lower, primarily-infected leaves to upper leaves with induced resistance. As treatment of healthy plants with low concentrations of ABA also reduced the numbers of lesions formed on subsequent challenge inoculation (Fraser, 1982), it is possible that ABA might have a direct effect on susceptibility, as opposed to an indirect effect related to water status.

The reduction in lesion size in acquired systemic resistance raises the question of whether this indicates reduction in virus multiplication per lesion, or merely a reduction in necrotic area. The evidence available is balanced equally between a reduction (Ross, 1966; Van Loon and Dijkstra, 1976) and no reduction in multiplication in resistant leaves (Balazs _et al_., 1977; Fraser, 1979; Coutts and Wagih, 1983).

One model to explain induced resistance suggests that it might represent an activated form of the primary localization mechanism (Ross and Israel, 1970). The evidence for involvement of wall thickening is contradictory. Simons and Ross (1971b) reported that phenylalanine ammonia lyase (PAL) activity (involved in synthesis of phenolics and lignin precursors) was the same in leaves with acquired systemic resistance as in controls, but that PAL activity in resistant leaves increased more quickly than in the controls, after challenge inoculation. They also suggested a more rapid oxidation of phenolic compounds to quinones which might inhibit virus multiplication. However, Fritig _et al_. (1973) found that PAL activity increased at the same rate after challenge inoculation of resistant and control leaves. They suggested that acquired systemic resistance could not be explained in terms of differences in PAL activity, although Van Loon (1982) has argued that differences in PAL activities could be shown if the results were expressed per resistant cell.

There have been two studies of the ultrastructure of control and acquired resistant tissue, but these have not dealt with changes in wall thickness or appearance. Fraser and Clay (1983) suggested that acquisition of resistance in tobacco was associated with an increase in the numbers of myelinic bodies - complex multilamellar structures between the cell wall and the plasmalemma or cytoplasm. Control tissue contained fewer, or more loosely-organized membranous structures. However, Appiano and D'Agostino (1985) reported that resistant and control tissue contained equal numbers of myelinic bodies, although these were in the vacuole or cytoplasm rather than adjacent to the walls. They also suggested that the myelinic bodies are predominantly lipid. Whatever their function, it seems unlikely that myelinic bodies are directly related to acquired resistance or pathogenesis-related proteins.

Experiments with a number of other enzymes which increase in activity during localization, such as peroxidase, polyphenoloxidase and catalase, have not shown any evidence for involvement in systemic acquired resistance (Batra and Kuhn, 1975; Stein and Loebenstein, 1976; Faccioli, 1979). One positive result, however, was the demonstration by De Laat and Van Loon (1983) that activity of the enzyme converting 1-amino cyclopropane-1-carboxylic acid to ethylene was higher in leaves with induced resistance. This caused a more rapid rise in ethylene production after challenge inoculation in resistant than in control leaves. Although the link between increased ethylene production and necrosis is well established (De Laat and Van Loon 1982; 1983), an effect on reduction in lesion size in leaves with acquired resistance remains to be proved.

Evidence for a second model; that reduction of lesion size in acquired resistance is separate from the primary localization response, is limited. Chessin (1982) showed that ultraviolet light increased the size of the primary lesions of TMV on pinto bean, but had no effect on acquired resistance, which may therefore have caused reduced lesion size in a separate, and ultraviolet-insensitive manner.

Stein et al. (1979) induced resistance to TMV in tobacco leaves by treatment with an ethylene maleic anhydride co-polymer. They suggested that the resistance was associated with a 'specific ribosome fraction', in that they found an increase in the tendency of ribosomes to aggregate during extraction and concentration. The nature of this 'stickiness' of ribosomes was not clear, nor was the relationship to resistance.

The final type of explanation for reduction in lesion size involves virus-induced changes in cytokinins. Application of exogenous cytokinins reduced necrosis and lesion size (Balazs et al., 1976), although more complex effects have been reported (reviewed by Fraser and Whenham, 1982). Sziraki et al. (1980) and Balazs et al. (1977) have presented evidence that necrotic TMV infection of half leaves of tobacco caused an increase in cytokinin concentration in the opposite, uninoculated half, and that this might limit necrosis on challenge inoculation. However, the cytokinin changes observed were small, and were entirely based on bioassay, which is not totally reliable for identification and accurate quantification.

6.6.3. Dark green islands

Systemic virus infection frequently causes a mosaic of light green and dark green areas. Generally, the dark green areas are free or comparatively free of virus (Loebenstein et al., 1977), tend to have higher chlorophyll concentrations than healthy leaves, and are to a greater or lesser extent resistant to challenge inoculation (Fulton, 1951). The biochemistry of mosaic symptom development is considered in Chapter 7, while the mechanism of resistance is considered here. There are two questions: how is the tissue resistant to invasion by the systemically-infecting virus, and how is it resistant to a second inoculation? These might involve the same or different mechanisms.

Loebenstein *et al.* (1977) showed that the dark green tissue of CMV-infected tobacco leaves was resistant to challenge inoculation by CMV, but not to TMV. The induced resistance was therefore virus-specific. At 30°C, resistance broke down and the dark green islands were invaded. In TMV-infected tobacco, there was some breakdown of the resistance of the dark green islands with time (Atkinson and Matthews, 1970), although this form of resistance could be long lasting. Thus Murakishi and Carlson (1976) were able to demonstrate some residual resistance in plants regenerated from protoplasts prepared from dark green areas.

Failure of virus to invade the dark green tissue was not apparently due to lack of contact, as Atkinson and Matthews (1970) found protoplasmic connections between cells in yellow and green areas. Their studies showed that the concentration of TMV particles fell away over a distance of several cells across the boundary, rather than abruptly. This might suggest that the resistance of dark green areas is caused by an inhibition of virus multiplication rather than by a barrier to transport. Some differences in protein composition of the light and dark areas have been reported, for example in Fraction 1 (Reid and Matthews, 1966) and in electrophoretic patterns (Kluge, 1976). But these offer no explanation of resistance mechanisms.

Sziraki and Balazs (1975) suggested that the dark green areas contained higher concentrations of cytokinins than light green areas. However, the differences were small, and based on bioassays. The reported effects of cytokinins in maintaining leaf chlorophyll concentration during senescence (e.g. Biddington and Thomas, 1978), and on susceptibility to virus infection (Fraser and Whenham, 1982) suggest that further examination of cytokinin metabolism using physical methods for assay and identification would be rewarding.

Whenham and Fraser (1985) and Whenham *et al.* (1986) have shown that ABA concentration is increased in dark green areas of TMV-infected tobacco leaves, and Fraser (1982) found that treatment of healthy plants with low concentrations of ABA decreased the number of lesions formed on challenge inoculation of a local lesion variety of tobacco with TMV. This suggests that a TMV-induced increase in ABA concentration in the dark green tissue might contribute to its resistance to inoculation. How the virus can alter hormone metabolism in non-invaded areas of leaf remains an interesting and unanswered question.

6.6.4. Cross protection

This phenomenon was first described by McKinney (1929), who showed that when tobacco plants were inoculated with a strain of TMV causing a green mosaic, no further symptoms developed when the plants were subsequently inoculated with a strain causing a yellow mosaic. Since then, a systemic infection by one strain has been shown to protect against challenge inoculation by other strains of the same virus, for many viruses and in many host species. Cross protection has been used extensively in practical crop protection, but in a limited number of hosts (e.g. Rast, 1975; Costa and Müller, 1980; Sequiera, 1984). For obvious commercial reasons, protecting strains which give mild symptoms and cause minimal loss of yield have been favoured. Whether the ability of these strains to protect is related to their reduced aggressiveness (attenuation) is an open question. The advantages and disadvantages of cross protection as a crop-protection strategy are considered in detail by Fraser (1985a). This section is concerned with the biochemical mechanisms which may be involved, and the theories which have been proposed to explain the effects. At the time of writing, some of the major questions remain unanswered.

It is worth stressing at the outset that the term 'cross protection' has been used to describe a number of effects, which might involve separate mechanisms. Thus it has been used in the context of the 'green island' effect described in section 6.6.3, where the 'protecting' virus is not present in the resistant tissue (Fulton, 1980), as well as where the presence of the protecting strain in the tissue is required for resistance (Cassells and Herrick, 1977a). Furthermore, a complex terminology has developed, with the terms 'interference', 'acquired immunity', 'antagonism', 'acquired tolerance', 'premunity' and 'cross-interference' being used to describe phenomena which may or may not be distinct.

To some extent, the probable existence of different types of mechanism of cross protection, even in an individual infected plant, has given rise to somewhat unnecessary argument (e.g. De Zoeten and Fulton, 1975; Zaitlin, 1976), with conclusions about one type of mechanism being tested and discarded, using evidence from another type. This is particularly relevant to the disputed role of coat protein considered below.

'Cross protection' has generally been used to describe interactions between related viruses, and has been used extensively to test for taxonomic relationships between pairs of viruses (Gibbs, 1969). However, it would be wrong to imply that cross protection is a universal effect. Absence of cross protection has been reported for RMV (Wilkins and Catherall, 1974) and MDMV (Paulsen and Sill, 1970) and other viruses. Even within a single host-virus combination, various levels of cross protection are possible. Thus Fulton (1978) tested eight isolates of TSV for cross protection in tobacco. Of the 28 possible combinations of strains, 12 pairs showed no cross protection when inoculated in either order; eight pairs protected reciprocally against each other; and eight pairs protected when inoculated in one order but not in the other. Any theory of the mechanism(s) of cross protection has to be able to explain such negative or unilateral effects as well as straightforward protection.

The mechanisms of cross protection may also overlap with interference between unrelated viruses in mixed or successive inoculations. These may lead to reduction in the pathogenic effects of one or both partners (Nitzany and Cohen, 1960; Nhu *et al.*, 1982), or to synergistic effects giving more serious disease or increased multiplication (Damirdagh and Ross, 1967; Dodds and Hamilton, 1972).

Diverse methods have been used to study cross protection, but most have given little quantitative information about its effects on the multiplication and spread of the challenging virus. Many investigators have simply recorded milder symptoms, or failure of the challenging strain to express a particular pattern of symptoms such as yellow mosaic. In cross protection in *N'*-gene tobaccos, use of a local lesion strain of the virus as the challenge inoculum for systemically infected plants allows estimation of the extent of cross protection by reduction in the numbers of lesions formed (De Zoeten and Fulton, 1975; Sherwood and Fulton, 1982). This type of protection and assay may refer more to events associated with the initial establishment of infection. In contrast, where the challenging virus also causes a systemic infection, the mechanism of protection could be directed against some later phase of virus multiplication. Assay methods used to estimate amounts of the protecting and challenging strain must be able to discriminate between the two, and have included differential local lesion assays (Cassells and Herrick, 1977a), strain-specific antisera (Cassells and Herrick, 1977b), and electrophoretic separation of strains with coat proteins differing in net charge (Dodds, 1982).

Mechanisms operating at the site of infection. If two isolates of the same virus are mixed and inoculated together, they may interfere, possibly as a result of competition for a limited number of infection sites. It has also been shown that coat protein can interfere with infection by the homologous virus, although very high concentrations were required for effect (Wu et al., 1962). In tobacco, Wu (1964) found that the VM strain of TMV reduced the number of U1 lesions formed in mixed inoculations. Ultraviolet irradiation of leaves immediately after mixed inoculation allowed U1 lesions to appear. This suggested that VM had blocked an early event in the establishment of U1 infections. Established infections could also block early infection events of challenging strains, and blocking of infection sites is only one possible mechanism. Sherwood and Fulton (1983) suggested that a TMV isolate causing a systemic infection of the N'-gene host Nicotiana sylvestris competed for infection sites used by a challenge inoculum of a local lesion strain, but that there was also interference with the multiplication of the challenge strain at another level.

The coat protein mechanism. For the TMV-N. sylvestris system, De

Zoeten and Fulton (1975) suggested that the RNA of the challenging virus, after uncoating to initiate its infection, might be sequestered in the coat protein of the protecting virus. Their argument was that the challenging strain RNA would be unlikely to be able to uncoat and initiate infection, when the cellular environment was already directed towards encapsidation and formation of progeny virus particles, and contained adequate or large amounts of coat protein.

However, the encapsidation theory was strongly criticized by Zaitlin (1976), who showed that cross protection could be induced in N. sylvestris by the PM-1 mutant of TMV, which has a defective coat protein. This criticism was reinforced when Sarkar and Smitamana (1981) found that a coat protein-free mutant could induce cross protection, and when Niblett et al. (1978) showed that viroids, which lack proteins, showed cross protection between strains.

Shalla and Petersen (1978) prepared protoplasts from leaves of N. sylvestris infected with a protecting strain of TMV, but were unable to detect virus antigen in many of them by fluorescent antibodies.

For viruses with multi-component genomes, where different strains differ in their abilities to cross protect, studies with pseudorecombinants have shown that control of the ability to cross protect may reside on a genomic segment other than that coding for the coat protein (Murant et al., 1968; Fulton, 1978; 1980).

These lines of evidence have been taken to show that the sequestration theory is untrue and that coat protein is not involved in cross protection. However, there are serious difficulties in interpretation. The work with defective or protein-less pathogens does show that a type of cross protection can be established in the absence of functional or any coat protein. But this conclusion does not exclude the possibility that coat protein might be involved in other mechanisms.

The need to consider more than one mechanism within the same plant is emphasized by the many studies of the N. sylvestris-TMV system. There can be little doubt that cross protection in the dark green areas is unlikely to involve coat protein, as little or no protecting virus is present. However, the situation might well be different in the light green, heavily infected tissue. Sherwood and Fulton (1982) compared the ability of the challenge strain and its RNA to infect protected plants, and concluded that cross protection in light green areas was a result of inability to uncoat. This might well be an effect of protecting strain coat protein. Pelcher et al. (1980) showed by in vitro experiments that heterologous coat protein can stabilize TMV particles to alkaline disassembly.

Some recent strong evidence for the involvement of coat protein comes from experiments with genetically-transformed tobacco plants (Abel et al., 1986). Plants in which a cloned DNA copy of the TMV coat protein gene was expressed were strongly protected against a second inoculation with the virus, whereas transformed plants not expressing the coat protein gene were fully susceptible.

Wilson and Watkins (1986) showed that virus coat protein subunits inhibited the cotranslational disassembly of TMV particles in vitro (see Chapter 2.2). The inhibition of translation was greater than when naked TMV RNA was used as template. The inhibition appeared to be specific for coat protein, as bovine serum albumin had little effect. Wilson and Watkins suggested that their results might provide an explanation of the mechanism of cross protection in vivo. The implication of these results, together with those of Abel et al. (1986) is that the presence of viral coat protein in cross-protected plants interferes with the very early stages of uncoating of any challenging virus. Conceivably, this could be at as early a stage as the initial uncoating of the 5' end to expose the first ribosome attachment site. This mechanism may also explain why protection is only against closely related viruses, in that the coat proteins must show sufficient similarity for uncoating to be inhibited.

The replication site-competition mechanism. This proposes that the protecting strain saturates some site or component required for virus replication, with the replicase as a favoured candidate (Gibbs, 1969; Barker and Harrison, 1978). Alternatively, Ziemiecki and Wood (1976) suggested that the protecting strain might alter the characteristics of binding of ribosomes to mRNA, to the detriment of translation of the challenging strain. As yet, there is little or no direct evidence for such mechanisms, and understanding will require an advance in our knowledge of the nature of viral replicases and how specificity of template recognition is determined.

Autoregulation. Kiho and Nishiguchi (1984) studied an attenuated strain of TMV, $L_{11}A$, which can protect tomato plants against subsequent infection by normal strains. The attenuated strain multiplied as well as the wild type for the first four days of infection, but less well thereafter. They named the effect 'autoregulation', and suggested that it might be due to an inhibition of viral RNA synthesis due to an overproduction of the 183 kDa viral protein. The implication is that the 183 kDa protein is involved in control of the relative amounts of synthesis of the different RNAs. Radiolabelling at four days after inoculation with $L_{11}A$ showed that synthesis of TMV RNA and the replicative intermediate were much less than with wild type, whereas synthesis of replicative form was similar to wild type.

The coat protein of $L_{11}A$ was shown to have the same amino acid composition as that of a mutant Ls1, which is defective in cell-to-cell movement at high growth temperatures (Nishiguchi et al., 1984). This result suggests, but does not prove, that coat protein is not involved in determination of either autoregulation or the transport function.

Nishiguchi et al. (1985) studied the base sequence of $L_{11}A$ and its ancestral strain L which multiplies normally. Of ten base substitutions in $L_{11}A$, seven occurred in the third bases of in-phase codons, and did not alter the amino acids specified. One of these substitutions did occur in or close to the sequence determining the initiation of assembly, where it might have had an effect separate from the coding property.

The remaining three substitutions occurred in the common reading frames of the 126 kDa and 183 kDa readthrough protein, and did cause amino acid changes. The changes are at positions 1117, 2349 and 2754.

The sequence of an intermediate and related attenuated strain, L_{11}, showed the change at position 1117 but not at the other two. The change at position 1117 therefore appears to be important in control of attenuation - although the complete sequence of L_{11} was not determined - but it remains to be fully established whether this is related to autoregulation and cross protection.

<u>Mechanisms based on minus-sense RNA</u>. Palukaitis and Zaitlin (1984) proposed that the (-) sense RNA of the challenging virus might be sequestered by the (+) sense RNA of the protecting strain. This model would certainly explain the ability of protein-less or protein-defective viroids and viruses to cross-protect. It also goes much of the way to explaining why related isolates tend to cross-protect whereas very dissimilar isolates do not; the ability to cross protect must be dependent on the degree of base sequence homology between the two isolates.

However, there are two problems with this model. Normally, the progeny (+) strand RNA is encapsidated in coat protein and only limited amounts might be available for hybridization with challenge strain (-) strand RNA. Secondly, why does the protecting strain (+) strand RNA not inhibit its own synthesis by binding to the protecting strain (-) RNA, with which it has 100% sequence complementarity? Palukaitis and Zaitlin argued that such feedback regulation did indeed occur in normally-infected plants, and there is evidence for accumulation of some double-stranded, full length viral RNAs late in infection (Jackson <u>et al</u>., 1971). However, the evidence for a feedback inhibition of viral RNA synthesis during normal virus accumulation is weak.

There is, however, evidence from studies with analogous systems that (-) sense RNAs can interfere with gene expression and viral infection. Izant and Weintraub (1985) showed that plasmid DNA containing actin gene sequences and directing transcription of the (-) sense strand could directly suppress expression of the actin genes of transformed chick cells. Coleman <u>et al</u>. (1985) prepared a plasmid which would produce large amounts of RNA complementary to bacteriophage mRNAs, and showed that these inhibited phage proliferation in the bacterial host. These studies suggest that antiviral mechanisms based on (-) sense RNAs are feasible; they do not have a direct bearing on whether 'natural' mechanisms of cross protection by plant viruses involve a (-) sense RNA.

6.7. CONCLUSIONS

It should be clear from this Chapter that the biochemistry of resistance has been intensively studied and yet in many areas it is still poorly understood. Perhaps the firmest conclusion that can be reached is that the mechanisms involved are diverse. One of the important general questions that remains is how increasing knowledge of the biochemistry of resistance and virulence might improve our ability to manipulate them in practical crop protection. There are several possible ways in which this knowledge might be deployed.

Knowledge of how resistance genes work, and in particular identification of the gene products, should help in isolation of the genes from genomic libraries. At present, with no gene products identified, such isolation would be immensely difficult and time-consuming. Some of the problems, and some possible alternative approaches for the future, are discussed by Hepburn et al. (1985).

Once clones containing resistance genes have been isolated, two major new avenues of exploitation are opened. Firstly, if coupled with suitable promoters, transformation vectors and plant regeneration systems, it will be possible to transfer resistance between unrelated species which suffer from the same virus. This should increase the capacity to utilize available genetic resources for virus control. Secondly, the resistance gene clones will be available for study and manipulation by recombinant DNA techniques. This should not only offer a quantum leap in our understanding of the biochemistry of resistance gene action, but also permit a widening of the genetic basis of resistance by probing for related sequences in other species, and by modifying existing sequences by site-directed mutagenesis.

These advances will be costly, and some possible lines of research and development will have to be assessed as much by the likely commercial benefits as by the intrinsic scientific interest. At this stage, the problem of viral mutations to virulence becomes particularly relevant. As with classical plant breeding - also a costly process - there is little long-term benefit to be gained from breeding a resistant cultivar, if there is a rapid response from the virus in the form of a resistance-breaking strain.

Knowledge of the biochemistry of resistance and virulence, and cloning of resistance genes, may offer solutions to this problem. Firstly, the ability to transfer genes between unrelated species, and to broaden the genetic base by site-directed mutagenesis, should make it possible to develop oligogenic resistance which may be more durable than the largely monogenic systems which are the primary concern of classical plant breeding. Secondly, further knowledge of the nature of virulence and resistance may help to identify types of mechanism with intrinsically high durability, as predicted in the models of host-virus interaction discussed in section 6.1.3 and in more detail elsewhere (Fraser, 1986; Fraser and Gerwitz, 1986).

The third type of possible advance in control of virus diseases comes not from knowedge of natural resistance mechanisms, but from understanding of the molecular biology of virus replication. This is now so far advanced for a few viruses, that it begins to be possible to think of ways of interfering in the plant by artificial means, by designing 'molecular spanners' to throw into the works of replication.

One such possibility is to use antimessengers: sequences which are complementary to viral messenger RNAs, which will interfere with translation, and for which model systems are already available (Izant and Weintraub, 1985; Coleman et al., 1985).

Another possible route is by use of satellites of plant viruses, which can regulate or inhibit the disease expression of the supporting virus (Waterworth et al., 1979). 'Synthetic' mixtures of CMV plus a disease-suppressing satellite have already been used as artificial inoculum, to control mosaic disease of Capsicum (Tine and

Chang, 1983). Baulcombe et al. (1986) have recently inserted a cDNA copy of CMV satellite RNA into the nuclear genome of tobacco. They demonstrated expression as an RNA, and presented indirect evidence that this could reduce the severity of CMV symptoms.

The assembly-initiation site of TMV can be isolated, and could theoretically be used as a competitive inhibitor of virus assembly. Many more types of 'molecular spanner' can be suggested; one based on the coat-protein model of cross protection may see early commercial development (Bialy and Klausner, 1986).

These futuristic types of control strategy have problems, of course. One is delivery: it does not make sense to attempt to deliver the active agent in effective quantities to individual plants, as would be done with a fungicide. The active agent has to be synthesized in the plant, either from some episomal form, or from DNA sequences integrated into the host genome. Another problem will be to ensure sufficient levels of expression to inhibit virus multiplication, without running the risk of pathogenic effects from the protecting agent. Also, there may be serious problems of specificity: mechanisms of resistance based on interactions between nucleic acids might recognize only closely related virus strains because of sequence variation. Mechanisms which rely on strongly conserved regions of the genome, or on protein-RNA interactions, might have a wider effective range. The problem of viral mutation to virulence will also have to be assessed, predictively, in this context. However, the optimistic outlook is that it should be possible to assemble a 'package' of these novel types of resistance agents which will give lasting protection against a spectrum of virus isolates.

CHAPTER 7

How do Viruses Control Plant Growth and Development, and Cause Symptoms?

7.1. INTRODUCTION: EFFECTS OF VIRUSES ON PLANT GROWTH AND DEVELOPMENT

Most virus infections inhibit growth. There are a few possible exceptions. Ghosh (1982) showed that RNMV increased growth parameters such as plant height, leaf size and stem diameter, in a number of host species although not in rice, where growth was inhibited. How the stimulation occurred is not certain, but it may be that localized virus infection can distort plant growth regulator balance in the remainder of the plant, with some stimulatory effects on growth. This explanation was suggested by Tuzun and Kuc (1985) for the stimulation of growth of tobacco injected with blue mould. Another possibility is specific damage to the vascular system resulting in altered partition of assimilate, with stimulatory effects on the growth of some organs.

 For growth studies of virus-infected plants, there are problems in assessing the extent of inhibition, and the mechanisms of control. The former might not appear to be a problem; after all, with the exception of cryptic infections (Lisa et al., 1981), the vast majority of virus-infected plants look diseased and smaller than healthy controls. But the difficulty lies in the many different ways of measuring inhibition of growth or loss of yield. The latter is directly relevant to agriculture, and the experimental complexities of assessment make it quite difficult to quantify how much loss of yield is caused by virus infections in different crops. 'Yield' is a complex term which can include a number of different parameters,

Table 7.1. Inhibition of growth or losses in yield caused by virus
infection.

Virus	Host	Infection	Percentage reduction	References
			Yield[a]	
various	wheat	natural	1	MacKenzie, 1983
BYDV	wheat	deliberate	6-79	Smith and Sward, 1982
BYMV/BCMV	bean	natural	53	Hampton, 1975
CPMV	cowpea	deliberate	56-85	Valverde et al., 1982
SCMV	maize	deliberate	28-46	Louie and Darrah, 1980
TMV	tomato	deliberate	0-67	Brisson et al., 1984
TMV	tomato	deliberate	10-20	Broadbent, 1964
			Growth	
PVY	tobacco	deliberate	10-39[b]	Latorre et al., 1984
TMV	tobacco	deliberate	5-60[c]	Fraser, 1972
TMV	tomato	deliberate	34-72[b]	Leal and Lastra, 1984
PNRV/PDV	peach	deliberate	60-93[b]	Smith and Neales, 1977b
BYDV	barley	deliberate	5-55[c]	Jensen, 1969
BYDV	wheat	deliberate	13-34[b]	Jensen, 1972

[a] seed or fruit. [b] shoot dry weight. [c] leaf fresh weight

economic as well as biological. These are discussed by Zadoks and
Schein (1979).

Table 7.1 shows examples of growth inhibition, or yield losses,
caused by representative virus infections. The range is highly
variable, but this is as much a consequence of differences in
experimental design as of differences in the inhibitory effects of
different viruses. Thus some of the data are from natural infections,
with a proportion of 'misses'. These data are representative of the

economic effects of a virus on crop productivity. Experiments where crops have been deliberately infected tend to give much higher losses, indicative of the maximum potential loss in an epidemic. The degree of inhibition of growth or yield in these experiments can, however, be markedly influenced by environmental factors (e.g. Chant, 1983; Helms et al., 1985); plant age at the time of inoculation (e.g. Agrios et al., 1985; Johnson et al., 1983); and host genetic background (e.g. Endo and Brown, 1963; Jones and Catherall, 1970).

The objectives of this Chapter are to consider how virus infections affect the primary processes of growth and development, and how they cause symptoms. The ways in which these effects might be inter-related will also be considered.

7.2. CONTROL OF GROWTH

7.2.1. Control models and mechanisms

Virus infection may control growth in the overall terms of the whole plant, or more specifically in relation to organs or tissues. These effects on growth rate are essentially quantitative, but they overlap with viral controls of morphogenesis, which may be largely qualitative.

There are two possible mechanisms for quantitative inhibition of growth: competitive and non-competitive. These make certain predictions about the relationships between certain parameters of infection.

The competitive model envisages virus multiplication as sufficiently active to create a significant sink for amino acids and nucleoside triphosphates. For example, in TMV infection of tobacco, the virus RNA and coat protein become by far the commonest types of nucleic acid and protein in the plant (see Figs 1.1 and 1.2). Diversion of precursors to form viral nucleoprotein, plus the attendant energy demands, might inhibit host protein and nucleic acid synthesis competitively. In this model, therefore, there should be a quantitative relationship between the amount of virus accumulating, and the extent of inhibition of plant growth.

The non-competitive model recognizes that for many viruses, the level of multiplication is very low in comparison with the host concentrations of nucleoproteins. Zaitlin (1979) pointed out that BYDV multiplication in barley is several orders of magnitude lower than TMV in tobacco, but severe symptoms and growth inhibition still result. Viruses such as BYDV are most unlikely to present any threat to host growth through competitive inhibition. In these cases, the inhibition of growth may be a consequence of the cytopathic effects of infection. One model is that the extent of growth inhibition might be related to the severity of symptoms caused. The model can also encompass more detailed experimental measurements, such as the effects of infection on photosynthesis and other growth-related metabolism.

Both models, being essentially quantitative, are open to mathematical analysis. Qualitative effects of virus infection on morphogenesis are likely to involve virus-induced changes in growth regulator metabolism. However, it is stressed that quantitative and qualitative effects can overlap; in particular, changes in quantitative parameters of growth might also be regulated at the hormonal level.

7.2.2. Experimental evidence for growth control mechanisms: quantitative effects

Investigation of quantitative relationships between growth, virus multiplication and symptom severity requires an experimental design which produces adequate variation in all three parameters. Variation is normally achieved by use of different virus isolates and different host lines. Resistance or 'tolerance' genes in the host can be exploited to increase variation.

Table 7.2 shows an analysis of interactions between these three parameters in TMV-infected tobacco (Whenham et al., 1985) and tomato (Fraser et al., 1986a). A total of 26 TMV isolates was used. These differed in intrinsic ability to multiply, and caused symptoms of varying severity, which were assessed on a scale of 0-3. One of the tomato hosts contained the Tm-1 gene for TMV resistance, which

Table 7.2. Ordinary (r) and partial (c) correlation coefficients for symptom severity, shoot fresh weight and virus concentration, in tobacco and tomato plants infected with various isolates of TMV.

	Fresh weight r	symptom score r	fresh weight c
Host: **tomato (+/+)**, 22 or 23 degrees of freedom			
Mean TMV concentration 1.78 mg; maximum 3.2 mg/g fresh weight			
Symptoms	-0.77 ***		-0.71 ***
Multiplication	-0.46 *	0.53 *	-0.10 (NS)
Host: **tomato (Tm-1/Tm-1)**, 21 or 22 degrees of freedom			
Mean TMV concentration 1.38 mg; maximum 3 mg/g fresh weight			
Symptoms	-0.82 ***		-0.73 ***
Multiplication	-0.60 **	0.58 **	-0.27 (NS)
Host: **tobacco**, 4 or 5 degrees of freedom			
Mean TMV concentration 4.2 mg; maximum 6.0 mg/g fresh weight			
Symptoms	-0.81 *		-0.65 (NS)
Multiplication	-0.86 *)	0.65 (NS)	-0.74 (NS)

*, ** and *** = statistically significant at P = 0.05, 0.01 and 0.001 respectively. NS = not significant at P = 0.05. From Fraser *et al.* (1986a).

reduced multiplication of some isolates of the virus, but still permitted systemic spread. Overall, TMV multiplied to a greater degree in tobacco plants.

Ordinary correlation coefficients (r) were calculated for the relationship between pairs of parameters, for each host. To eliminate effects of interactions between parameters, partial correlation coefficients (c) were also calculated, from multiple regression analysis. Note that the tobacco experiment involved a smaller number of TMV isolates than the tomato experiments. Consequently, higher values of the correlation coefficients were required for any given level of statistical significance in the tobacco experiment.

Considering first the ordinary correlation coefficients, the results show that plant growth was more strongly correlated with symptom severity than with TMV multiplication in the two tomato hosts, although all correlations were statistically significant. In tobacco, both r values were high and significant. TMV concentration was significantly correlated with symptom severity in the two tomato experiments, but not in the tobacco experiment.

The partial correlation coefficients for the relationship between symptom severity and growth in the two tomato hosts were still strong; almost as high as the corresponding r values. In contrast, the c values for the relationship between TMV concentration and growth were much lower than the r values, and were not statistically significant. These results show clearly that plant growth was much more related to symptom severity than to TMV multiplication in tomato. The apparent relationship between multiplication and growth suggested by the simple correlation coefficients was a result of an interaction between multiplication and symptom severity.

In tobacco, partial correlation coefficients for the relationship between TMV concentration or symptom severity, and growth, were both slightly lower than the simple correlation coefficients, suggesting only a small interaction between symptom severity and TMV multiplication. However, it is clear that both symptom severity and TMV multiplication inhibited growth directly, and were of approximately equal importance.

The relative importance of the different parameters revealed by c values was further examined by using them to calculate the percentages of observed variation in growth which could be accounted for by different influences: these are shown in Table 7.3. Firstly, each host had a level of intrinsic variation which would have occurred without any experimental treatment. Of the remainder, most

Table 7.3. Multiple regression analysis attributing variation in the growth of TMV-infected plants to various sources.

Host	Factor accounting for variation	Percentage of variation accounted for
Tomato (+/+)	Symptom severity TMV multiplication	55
	Plant - intrinsic	9
	Unattributed	36
		100
Tomato (Tm-1/Tm-1)	Symptom severity TMV multiplication	60
	Plant - intrinsic	15
	Unattributed	25
		100
Tobacco	Symptom severity TMV multiplication	76
	Plant - intrinsic	10
	Unattributed	14
		100

From Fraser _et al_. (1986a).

could be accounted for by variation in symptom severity plus variation in TMV multiplication; only comparatively small percentages of variation in growth could not be accounted for in each of the hosts. This result suggests that the input parameters in these analyses were realistic, and contained a sufficiently comprehensive expression of the pathogenic process, to explain most of the effects on growth. In the case of tomato, the multiple regression analysis (Table 7.3) suggested that only a small proportion of the 60% of variation attributable to symptoms and TMV multiplication was directly attributable to the latter. In tobacco, TMV multiplication and symptom severity contributed about equally to the 76% of variation in growth accounted for.

The importance of TMV multiplication in direct control of growth in tobacco, but not tomato, might be a result of the higher concentrations of virus reached in tobacco. If this is correct, it suggests that competitive inhibition of plant growth does occur, but that the mechanism only operates at the top end of the scale of virus multiplication. Even with the TMV levels reached in the tomato experiments - which are still very high by comparison with most plant viruses - it appeared that the sizes of the sinks for nucleoprotein precursors and energy offered by the virus isolates were not enough to cause a serious competitive inhibition of host growth. It is therefore necessary to seek an explanation of the inhibition of host growth based on other factors, which are, however, related to symptom severity. Some possible mechanisms are considered in sections 7.2.3 and 7.2.4.

Other investigations of the quantitative relationships between symptom severity, growth or yield, and virus multiplication, have been less complete. Good correlations between symptom severity and reduction in yield have been found in Lolium infected with RMV (Wilkins and Catherall, 1974), soybean with BPMV (Windham and Ross, 1985), oats with BYDV (Endo and Brown, 1963), and wheat with WSMV (Johns et al., 1981). In contrast, Kingsland (1980) found that the disease severity of individual maize plants infected with MDMV did not correlate well with yield loss.

Differences in virus multiplication were not investigated in any of these experiments. Scott (1982) found differences in WCMV multiplication in clover cultivars, but these were not correlated with differences in yield loss.

7.2.3. Growth inhibition, photosynthesis and respiration

Development of visible symptoms suggests changes in the chloroplasts, and changes in chloroplast ultrastructure have been reported for many virus infections (see Chapter 5.2.3). Some authors have therefore looked to a reduction in the rate of photosynthesis for an explanation of inhibition of growth. Others have also examined changes in respiration and carbohydrate levels for a more complete picture of changes in energy flow, and its relation to growth.

For tomato plants infected with ToYMV, Leal and Lastra (1984) found reductions in photosynthetic rate and an increase in respiration, and suggested that these changes might explain the reduced growth of infected plants. In contrast, most other authors have concluded that inhibited photosynthesis does not provide a direct explanation of stunting. Thus in peach with PRD, the maximum inhibition of photosynthetic rate (Smith and Neales, 1977a) was much smaller than the inhibition of growth (Smith and Neales, 1977b). In wheat and barley infected with BYDV, Jensen (1968; 1972) found inhibition of photosynthesis and stimulation of respiration of sufficient magnitude to account for the observed inhibition of plant growth. But both infected hosts contained grossly elevated levels of sugars and starch, so photosynthate and energy supply were most unlikely to have limited growth. Jensen (1969a) suggested that the observed changes in respiration and photosynthesis may have reflected feedback controls on these processes by the high concentrations of sugars.

The effects of virus infection on photosynthesis and respiration, as metabolic processes in their own right, are considered in Chapter 5.

7.2.4. Qualitative effects of infection: morphogenesis

Morphogenesis is considered in this section in terms of gross changes in organ formation or plant development. Morphogenetic processes on a finer scale are involved in some aspects of development of visible symptoms, and are considered in that light in sections 7.3.1 and 7.3.2.

Several authors have recorded major morphogenetic or developmental effects after infection, and have proposed explanations based on changes in plant growth regulators. Flores and Rodriguez (1981) noted that CEV impaired the ability of Gynura cuttings to form roots, and suggested that viroid-induced changes in an auxin-like substance were involved. Bailiss and Senananyake (1984) found that EAMV and BYMV infection of faba beans delayed senescence, and increased branching and the numbers of inflorescences produced. Van Steveninck (1959) recorded various effects of PMV on pod abscission of lupins, and proposed explanations based on altered concentrations of endogenous growth substances including auxins. Several plant viruses cause premature abscission of infected leaves: this may be due to virus-induced changes in auxin concentration (Dyson and Chessin, 1961), but is likely also to involve changes in ethylene. These are also involved in virus-induced epinasty (Levy and Marco, 1976).

In an interesting report by Fraser and Matthews (1981), inoculation of cotyledons of Chinese cabbage seedlings with TYMV was shown to cause a transient retardation of the rate of leaf initiation at the apex. The inhibitory factor moved out of the cotyledons within a few hours of inoculation, and preceded the spread of the virus by several days. Application of ABA to cotyledons of healthy plants produced similar effects, suggesting that ABA is the 'messenger' in virus-infected plants (Fraser and Matthews, 1983).

The leaves of healthy plants may display a characteristic 'growth rhythm', moving through a helical path. Novak (1975) showed that TMV-infection of tobacco disorganized this severely, and under some circumstances could induce unusually active movement. The mechanism remains unclear.

7.2.5. Effects of infection on plant growth regulators

Much remains to be discovered about the metabolism and functions of plant growth regulators. There is lively and continuing discussion (e.g. Trewavas, 1981) of whether they can be considered as hormones, in the sense used in studies of animals, and about how they act biochemically to bring about changes in growth and development.

The objective of this section is to review the ways in which virus infections can alter growth regulator metabolism, and to question whether this tells us anything further about the control of growth and development of the infected plant. A more comprehensive review of the literature on changes in growth regulator concentrations in virus-infected plants is given by Fraser and Whenham (1982). They also considered the numerous reports of effects of exogenous growth regulators on different types of virus infection, and concluded that such studies have given disappointingly little insight into plant-virus interactions. This type of experiment is only considered here if it attempted to mimic or make good the observed effects of infection on the concentrations of endogenous growth substances.

Identification and assay of growth substances in the plant is difficult. Most occur at very low concentrations, and in a variety of related molecular forms. Early experiments relied extensively on bioassays, which are not always fully reliable for identification or quantification. Modern physical methods based on gas chromatography, high performance liquid chromatography, and mass spectrometry are available, but must be accompanied by methods to correct for losses during the extensive pre-purification normally required. New methods for radio-immunoassay of plant growth regulators also offer an increase in specificity of response, especially when based on monoclonal antibodies.

Auxins. Most reports have associated the stunting of virus-infected plants with reduced auxin levels. Thus Smith *et al.* (1968) found that BCTV reduced auxin levels in three different hosts. Rajagopal (1977) found that a systemic infection by TMV reduced the concentrations of phenylacetic acid and indolylacetic acid (IAA) by as much as 95% in tobacco, while increasing the concentrations of their precursors phenylalanine and tryptophan. The changes occurred very soon after infection, when most of the leaf cells should still have been uninfected. There do not appear to have been any attempts to reverse virus-induced stunting by treatment of plants with additional auxins.

Among possible mechanisms for decrease in auxin concentration, Lockhart and Semancik (1970) noted an increase in IAA oxidase activity in CPMV-infected cowpea. Several authors have reported increased peroxidase activity after infection (see Chapter 4.4.2); this enzyme has IAA oxidase activity (Endo, 1968).

Other workers have found auxin concentrations unchanged by infection (e.g. of barley by BYDV; Russell and Kimmins, 1971), or have actually recorded increases (e.g. in tobacco reacting hypersensitively to TMV; Van Loon and Berbee, 1978).

Abscisic acid (ABA). ABA is normally synthesized in response to water stress, then quickly metabolized to inactive forms when the stress is removed. There are various 'detoxification' pathways for removal of

Table 7.4. Effects of tobacco mosaic virus infection on the concentrations of abscisic (ABA) and phaseic (PA) acids in tomato and tobacco.

Host	TMV isolate[a]	ABA (ng/g)	PA (ng/g)
Tobacco	healthy	30 ± 4	745 ± 86
	vulgare	82 ± 9	2436 ± 375
	flavum	280 ± 94	nd
Tomato	healthy	266 ± 10	124 ± 32
	MII-16	284 ± 23	62 ± 9
	vulgare	338 ± 11	54 ± 7

[a]*Vulgare* caused severe systemic mosaic and MII-16 caused mild mosaic. *Flavum* caused necrotic local lesions. ABA and PA concentrations were measured 15 days after inoculation. Values are means ± standard errors. nd = not determined.

this growth inhibitor (reviewed by Zeevaart and Boyer, 1982). They include formation of the ABA-glycosyl ester and an oxidative pathway to phaseic acid (PA) and dihydrophaseic acid (DPA). Zeevaart and Boyer concluded that the pathways of ABA metabolism varied between species, and that the metabolites formed were stable. However, most studies of virus-infected plants have assayed only the free acid, and have neglected the metabolites.

The reported effects of virus infection on abscisic acid (ABA) metabolism have been rather varied. With the exception of one report of a considerable reduction in ABA concentration in the early stages of infection of tobacco by TMV (Rajagopal, 1977), most workers have found unchanged or increased ABA, for example in CMV-infected cucumber (Bailiss, 1977; Aharoni *et al*., 1977).

Table 7.4 shows changes in TMV-infected tobacco and tomato. A local lesion infection of tobacco caused a greater increase than a systemic infection. The increase associated with lesions began to occur at about the time of lesion appearance. ABA was shown to be transported from the hypersensitively-reacting leaves to uninfected upper leaves in sufficient quantity to explain the observed inhibition of their growth (Whenham and Fraser, 1981).

In systemically-infected tobacco, although the extent of increase in ABA was smaller, there was a large increase in the absolute concentration of PA. The results in Table 7.4 indicate that the flux through the ABA metabolic pathway was increased by about 115 ng/g/day as a result of systemic infection. This must indicate that systemic TMV infection stimulated ABA synthesis more than it increased the actual ABA concentration.

The increase in ABA caused by systemic infection was, however, shown to be sufficient to explain much of the inhibition of growth of infected plants, by experiments in which healthy plants were sprayed daily with low concentrations of ABA (Whenham and Fraser, 1981). This was enough to maintain the average concentration of free acid ABA at the same level as in infected leaves, despite continuous metabolism to PA. The amounts of growth inhibition caused by TMV infection and ABA treatment of healthy plants were very similar. Exogenous ABA and TMV infection were also shown to inhibit growth by reducing cell division but not cell expansion.

Whenham et al. (1985) infected tobacco with a number of isolates of TMV which caused systemic symptoms of varying severity, and which multiplied to different extents. The results (Fig. 7.1) show that the increase in ABA concentration after infection was strongly correlated with symptom severity, but only weakly correlated with virus concentration. ABA increase and virus multiplication were both strongly correlated with the inhibition of growth of infected plants. In a multiple regression analysis of the type shown in Table 7.3, approximately 35% of the variation in growth of plants could be attributed to variation in ABA concentration and about 40% to variation in TMV multiplication.

These results suggest that the mechanism by which virus infection stimulated ABA synthesis might be related to the mechanism of production of visible symptoms; damage to the chloroplasts would be a

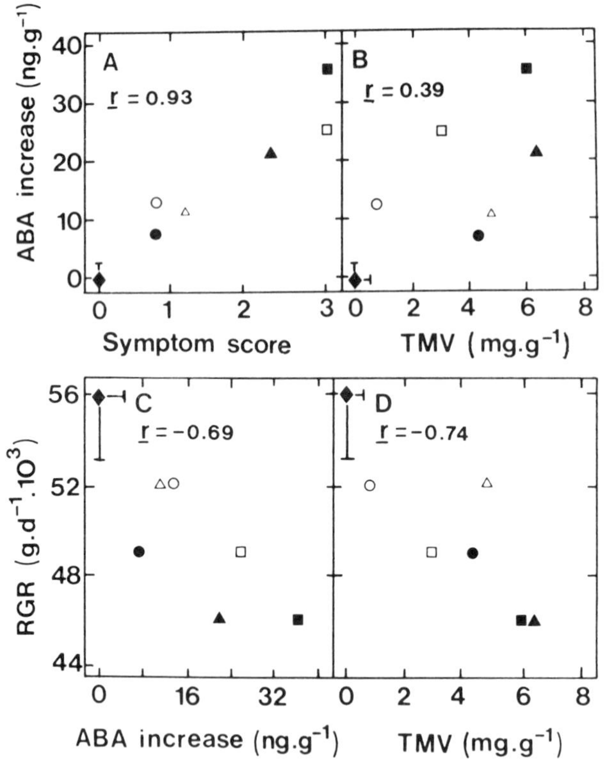

Fig. 7.1. Changes in abscisic acid concentration in TMV-infected tobacco leaves, and relationships between abscisic acid metabolism, virus multiplication and shoot growth rate. Plants were inoculated with TMV isolates causing mosaic symptoms of varying severity. (◆), healthy (sham-inoculated) control. TMV strains in order of increasing symptom severity: (○), MII-16; (●), V4; (△), N2; (▲), vulgare; (□), flavum; (■), U5. The bars on the healthy plant points indicate least signficant differences at P = 0.05. (A). Relationship between increase in ABA concentration and mosaic symptom severity. The ABA concentration in healthy plants was 9.8 ng/g. (B). Relationship between increase in ABA concentration and virus multiplication. (C). Relationship between shoot relative growth rate (RGR) and increase in ABA concentration caused by infection. (D). Relationship between shoot relative growth rate and virus multiplication. From Whenham et al. (1985).

likely common feature. ABA is known to be sequestered in the chloroplasts of unstressed healthy plants (Loveys, 1977), but there is still argument about whether it is synthesized within the chloroplasts (Milborrow, 1974) or outside them (Hartung et al., 1981). Whenham et al. (1985) showed that after TMV infection, the increase in ABA concentration was almost entirely extra-chloroplastic; the concentration of ABA within the chloroplast was not significantly altered by infection. This result tends to suggest that the receptor for growth inhibition by ABA is extrachloroplastic.

The relationship between symptom severity, chloroplasts and stimulation of ABA synthesis therefore remains to be clarified. It is established that virus-stimulated changes in ABA synthesis and transport play an important role in the control of growth of TMV-infected tobacco.

This conclusion cannot, however, be applied to other plants and viruses. In tomato (Table 7.4), TMV infection did not increase ABA concentration, and did not result in an increase in flux through the detoxification pathways. One point of interest is that overall ABA concentrations in healthy, unstressed tomato were 10-20 times those in comparable tobacco plants; the other is that exogenous ABA had little effect on the growth of healthy tomato plants even when applied at very high concentrations (Fraser et al., 1986b).

These differences in ABA metabolism and response are perhaps not surprising given the differences in growth habit between tobacco (strong apical dominance and monocarpic senescence) and tomato (weak apical dominance, and continuous production of inflorescences). The differences in effects on ABA do warn against seeking a single explanation of the effects of virus infection on plant hormone metabolism, and simple explanations for the inhibition of plant growth by viruses.

The possibility that the increase in ABA concentration caused by TMV infection might be involved in the higher rates of RNA synthesis found in infected leaves was considered in Chapter 3.2.1.

Cytokinins. Many viruses alter the apparent patterns of leaf senescence, and many do not invade juvenile tissue such as meristems and embryos. These results suggest an interaction with host cytokinin metabolism. This is supported by the fact that symptom formation and virus multiplication can be influenced by the application of compounds with cytokinin activity (Fraser and Whenham, 1978b). However, there have been very few studies of the effects of infection on metabolism of endogenous cytokinins, and all of these have used bioassays.

Cytokinin activity was reduced in TRSV-infected cowpea (Kuriger and Agrios, 1977) and *Nicotiana glutinosa* (Tavantzis *et al*., 1979). In contrast, increased cytokinin activities were reported in TMV-infected tobacco (Sziraki *et al*., 1980). It was suggested there that the increased cytokinin activity in uninoculated parts of plants with a local lesion infection elsewhere was involved in the mechanisms of acquired systemic resistance (see Chapter 6.6). For systemic infections, Sziraki and Gaborjanyi (1974) suggested that cytokinin activities were higher in the dark green, virus-free regions of mosaic. In BGMV-infected beans, De Fazio (1981) suggested that the higher cytokinin activities observed were related to the delay in leaf senescence caused by infection.

Sziraki and Balazs (1979) suggested that TMV RNA itself contained a large number of residues with cytokinin activity. Bioassays of fractions from an enzymic hydrolysate appeared to show activity corresponding to zeatin and isopentenyladenosine. However, re-examination using a highly selective and sensitive gas chromatographic method for cytokinins failed to detect measurable levels of any known cytokinin in TMV RNA (Whenham and Fraser, 1982).

Ethylene. Increased ethylene production has been reported mainly for infections which cause necrotic reactions, for example in tobacco leaves inoculated with TMV (Pritchard and Ross, 1975) or TNV (Gaborjanyi *et al*., 1971; De Laat and Van Loon, 1983). In a careful study of the relative times of changes in different stages of ethylene synthesis, De Laat and Van Loon (1982) showed that there was

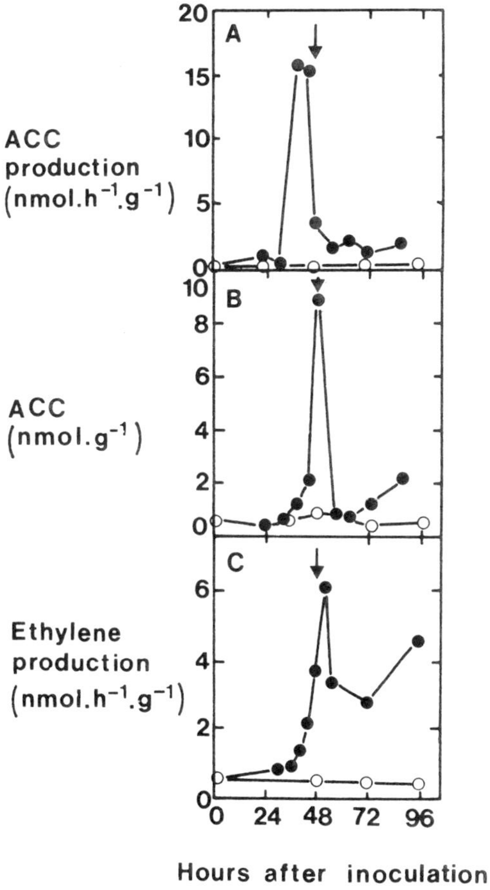

Fig. 7.2. Time courses of the production of ethylene and metabolism of its precursor ACC in tobacco leaves reacting hypersensitively to inoculation with TMV. (A). Aminocyclopropane carboxylic acid (ACC) synthase activity, measured as the amount of ACC accumulated in leaf disks incubated under nitrogen for 4 h. (B). ACC concentration in tobacco leaves at various times after inoculation. (C). Rate of ethylene production. In each case, (●) indicates TMV-infected leaves; (↓) indicates the time of lesion appearance, and (○) indicates water-inoculated healthy controls. From De Laat and Van Loon (1982), by permission of the authors and the American Society for Plant Physiology.

a large increase in the production of the immediate precursor of ethylene, 1-amino cyclopropane 1-carboxylic acid (ACC), and of ACC synthase activity, just before lesion appearance (Fig. 7.2). Ethylene synthesis was at a maximum slightly later, at the time of lesion appearance, and there was a further rise as lesions increased in diameter. De Laat and Van Loon (1982) concluded that ethylene production was regulated mainly at the level of ACC production. The earlier precursors in the pathway, methionine and S-adenosyl methionine (De Laat et al., 1981), were available but their concentrations did not alter pathway flux. These comprehensive studies are a good example of how virus-infected plants have given useful fundamental information about the metabolism of plant growth regulators.

Ethylene production has also been reported to be increased in cucumber cotyledons developing chlorotic lesions after infection with CMV (Marco and Levy, 1979). Production increased as chlorophyll concentration fell. Most studies of systemic virus infections, however, have found no change in ethylene synthesis (Nakagaki et al., 1970; De Laat and Van Loon, 1983). Further examples of ethylene changes in different types of infection are given by Fraser and Whenham (1982).

The trigger for the stimulation of ethylene synthesis is not yet known, but it would appear from the association with necrotic and chlorotic infections, and from the timing relative to development of visible symptoms, that it may be a result of early effects of wounding. It is interesting that ethylene production is thought to occur at the plasmalemma (Lieberman, 1979); among the early effects of necrotic infections are increases in membrane permeability and electrolyte leakage (Chapter 6.4). The evidence bearing on whether increased ethylene production is involved in virus localization and the necrotic local-lesion response is also considered there.

Ethylene production as a result of necrotic infection has been shown quite conclusively to be responsible for the marked epinasty of leaves in certain species (Ross and Williamson, 1951; Levy and Marco, 1976). Marco et al. (1976) also suggested that increased ethylene production was involved in the inhibition of hypocotyl elongation of CMV-infected cucumber seedlings, but was not the only factor.

Gibberellins. Most interest has centered on whether virus-induced changes in gibberellin concentrations can be responsible for the inhibition of stem elongation seen in many infections. Virus-induced stunting has been associated with reduction in gibberellin concentration in some hosts, for example in barley with BYDV (Russell and Kimmins, 1971), although in others such as TAV in tomato, infection caused no change (Bailiss, 1968). Several investigators have tried to reverse virus-induced stunting by application of exogenous gibberellins, with partial success in some host-virus combinations but not in others (reviewed by Fraser and Whenham, 1982).

In cucumber seedlings infected with CMV, hypocotyl elongation was inhibited. Bailiss (1974) found that this was associated with a reduction in the concentrations of endogenous gibberellins, but there were no qualitative differences between gibberellins of healthy and infected plants. In contrast, Ben-Tal and Marco (1980) found qualitative changes in gibberellins in the same system. They suggested that CMV may have altered the pattern of gibberellin degradation. Aharoni et al. (1977) were unable to assess the importance of the reduction in gibberellin concentration in the inhibition of growth of infected hypocotyls, because ABA concentration was also increased by infection, and ethylene too has been implicated as discussed earlier. Treatment of healthy seedlings with exogenous gibberellins increased hypocotyl growth, whereas treatment of infected plants did not (Fernandez and Gaborjanyi, 1976). Exogenous ABA inhibited the growth of infected hypocotyls, but a much higher concentration was required to inhibit the growth of infected hypocotyls. These results may mean that growth was under a dual control involving ABA and gibberellins, and that one effect of CMV infection was to alter the sensitivity of the tissue to both substances.

Other potential growth regulators. Polyamines are now recognized as a class of plant growth regulators, and have been particularly related to stress conditions (Altman et al., 1982). Altered polyamine metabolism has been reported in Chinese cabbage leaves infected with TYMV (Balint and Cohen, 1985). However, this has been considered primarily in relation to the polyamines present in the TYMV particle (see Chapter 2.5.1), and the possible effects on growth in this and other host-virus combinations have not been examined.

Oligosaccharins, perhaps derived from cell walls, are increasingly being recognized as potential growth regulators. Their chemical structure imparts a potential for information content or specificity which could make them precise regulators of morphogenesis (Van et al., 1985). To date, they have not been reported in control of growth and morphogenesis of virus-infected plants. It may be relevant, however, that cellulase digests of cell walls from tobacco leaves have been shown to impart an effect resembling acquired systemic resistance to TMV, when injected into leaves of a hypersensitive tobacco cultivar (Modderman et al., 1985).

7.3. HOW SYMPTOMS ARE FORMED

7.3.1. The diversity of symptoms

Symptoms are visible or otherwise detectable expressions of disease. Bos (1978) gave a comprehensive description of the different types of symptoms associated with virus diseases of many species, and of the use of symptoms in diagnosis of viral diseases. The effects of viruses on host metabolism, growth and development which have been described earlier in this book also fall within the definition of symptoms. So too do the visible manifestations of resistance mechanisms, such as the hypersensitive response. We are concerned here with a residual group of phenomena not previously considered in other contexts.

The preceding discussion has assumed that virus disease symptoms are easily recognizable, but there are some exceptions. Some viruses form 'cryptic' infections with no visible manifestation (e.g. Lisa et al., 1981); others may appear symptomless, but growth analysis can show inhibition. Some plants which may appear 'healthy' are in fact universally infected by a virus or viruses, and the healthy plant 'baseline' is not revealed until the viruses are eliminated by tissue culture techniques and thermotherapy (Walkey et al., 1982; Cassells and Minas, 1983).

Symptoms are very diverse. They may be the results of mechanisms operating at the tissue, cell or molecular levels. It follows that there may be several pathways for symptom formation. Some types of symptom, such as inhibition of growth, can be expressed in precise quantitative terms. Others, for example the severity of visible mosaic, are more difficult to quantify, and have to be assessed subjectively. Hebert (1982) has drawn attention to the different types of relationship which can exist between the eye's perception of disease severity and the intensity of the stimulus. For assessment of disease symptoms, visual acuity can be proportional to the logarithm of the stimulus (the Weber-Fechner law). Sufficiently quantitative data can be extracted, with transformations if necessary, for statistical analysis. Finally, there are symptoms which can best or only be described in qualitative terms.

7.3.2. Mechanisms involved in mosaic formation

The most striking symptom of many systemic virus infections is the development of a mosaic, usually of light green and dark green areas, on infected leaves. What determines the severity of mosaic, and what causes the pattern?

Zaitlin (1979) has pointed out that two different viruses may multiply to very varying extents, and yet cause equally severe symptoms. He noted, however, that for any one virus, there could be a relationship between multiplication and symptom severity. Fig. 7.3 shows such a relationship for TMV-infected tomato. The correlation of mosaic symptom severity with virus multiplication was statistically significant. The correlation matrix in Table 7.2 gives further examples. However, the data derived for the correlation in Fig. 7.3 indicate that variation in TMV multiplication could only account for

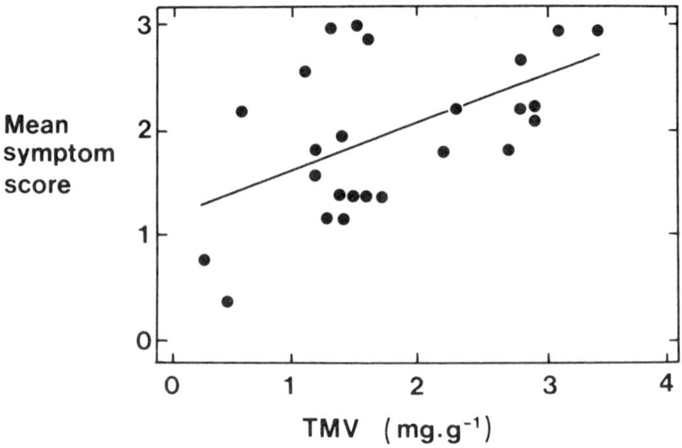

Fig. 7.3. Relationship between TMV multiplication and severity of mosaic symptoms in tomato. Plants were inoculated with 25 distinct isolates of the virus. From Fraser et al. (1986a).

28% of the variation in symptom severity on a simple correlation analysis. The corollary is that factors intrinsic to different virus isolates, other than ability to multiply, were responsible for the remaining 72% of variation in mosaic symptom severity.

Relationships between virus content and the severity of visible symptoms have also been shown for BYDV-infected oats (Jedlinski et al., 1977), and wheat and barley (Skaria et al., 1985). However, the magnitude of the effect of multiplication on symptom severity was not analyzed mathematically.

Zaitlin (1979) suggested that for TMV-infected tobacco, the effects of virus multiplication and effects on mosaic symptom formation could be separated by experimental treatment. He cited the evidence of Tomlinson et al. (1976) that methyl benzimidazole-2yl-carbamate (MBC; carbendazim) prevented mosaic symptoms without inhibiting virus multiplication. However, their estimates of virus concentration were based on a small number of infectivity assays only. Other studies showed that virus multiplication was reduced by up to 95% by MBC, during the period when symptoms were forming (Fraser and Whenham,

1978b).

There have been various attempts to explain the ontogeny of light green/dark green mosaics. Virus-induced changes in ABA and cytokinin metabolism have been proposed as explanations of the resistance of the dark green areas to challenge inoculation with virus, and there is also evidence that these changes in growth regulators may contribute to mosaic formation. The evidence is considered in the context of the induced resistance in Chapter 6.6.3. On balance, there appears to be strong evidence that virus-induced changes in ABA are involved in the formation of dark green areas and in their resistance to infection. Whenham et al. (1986) showed that the increase in ABA concentration caused by TMV infection of tobacco leaves was greater in the dark green (virus-free) tissue than in light green (infected) areas. Furthermore, treatment of healthy leaves with physiologically reasonable doses of ABA also caused them to turn darker green. The implication is that the TMV-stimulated increase in ABA concentration in infected leaves may cause the dark green areas, although how the virus can cause this in virtually uninfected regions requires explanation. The experimental evidence for the involvement of cytokinins in formation of dark green areas is more indirect.

A clearly-defined mosaic of light green and dark green areas is only formed in leaves which become infected by systemic spread of virus, whilst still at a very early stage in their development. This led Reid and Matthews (1966) to suggest that each island of dark green tissue in TYMV-infected Chinese cabbage was formed by division of a single cell or small group of cells which had become resistant, and conferred resistance on all its progeny. To a limited extent, this 'clonal' theory of the origin of dark green islands is supported by the evidence that plants regenerated from dark green tissue of TMV-infected tobacco leaves were transiently resistant to infection by TMV.

Carlson and Murakishi (1978) applied a novel test of the clonal origin theory. TMV-infected tobacco plants heterozygous for the Su (sulfur) phenotypic marker produced genetically-marked green (+/+) or yellow (Su/Su) spots on a yellow-green (Su/+) background by somatic recombination. The cells of each marked area were assumed to have arisen from a single recombination event, and were thus clonally related. In leaf ontogeny, the upper epidermis plus mesophyll is derived from a different germ layer of the apical meristem from the lower epidermis. The phenotypic marker was used to identify areas where upper and lower tissues were derived from different clones, but where the whole area was contained within a TMV-induced green island. Plants were regenerated from separated upper and lower tissues of the genetically-marked spot, and indexed for virus. Knowing the proportion of TMV-induced dark green area on the leaf (7.5%), it was calculated that only 0.56% (0.075^2) of the regenerated pairs from genetically-marked spots should have been virus-free in both tissues, if the virus-induced dark green tissue also had a clonal origin. In fact, 38% of genetically-marked spots were virus-free in the regeneration tests. This clear result led Carlson and Murakishi (1978) to conclude that TMV-induced dark green islands did not have a clonal origin, and that some diffusable substance must have been involved.

Green islands also form around the infection sites of many microbial parasites. Dekhuijzen (1976) has reviewed the evidence that increased cytokinin levels are involved in their formation.

7.3.3. Symptom formation at the cell and organelle level

The complex effects of different isolates of TYMV on chloroplast ultrastructure (Ushiyama and Matthews, 1970; Hatta and Matthews, 1974) and on 'sickling' (Matthews and Sarkar, 1976) and clumping (Chalcroft and Matthews, 1967b) have been related to the development of macroscopic mosaic symptoms (Matthews, 1973). The various pathways of symptom formation are doubtless related to the replication of TYMV in vesicles in the chloroplast outer membrane (Mouches et al., 1974). Fraser and Matthews (1979) showed that the isolate of TYMV controlled whether the chloroplasts fragmented or underwent the 'sickling' response. Both processes were dependent on photosynthesis because they required illumination, and both were inhibited by an inhibitor of photosynthetic electron transport.

With TMV, Jockush and Jockusch (1968) have provided a convincing explanation for how one class of virus isolates causes a particular type of symptom. Isolates producing yellow-green mosaic, such as flavum, were shown to have proteins which were temperature-sensitive in vitro. (Jockusch, 1966a). The coat proteins of almost all of these isolates were shown to be one charge less negative than that of the vulgare wild type, by electrophoresis on non-denaturing gels in the presence of 8M urea. These properties were correlated with accumulation of denatured coat protein in an insoluble form in plants grown at the restrictive temperature (Jockusch, 1966b).

Jockusch and Jockusch (1968) argued that this precipitated protein would lead to disruption of organellar structure. There is supporting evidence for particularly early degradative effects on chloroplast components after infection with strain flavum (Fraser, 1969). Peterson and McKinney (1938) also showed that when tobacco was infected with strains of a virus now presumed to be TMV, and causing different types of mosaic, the increase in chlorophyllase activity was particularly great with a yellow mosaic strain.

Overall, the correlation of the alterations in coat protein properties with yellow mosaic symptoms was very strong for a large number of temperature-sensitive TMV isolates and mutants (Jockusch, 1966a; 1966b). One exception is the mutant Ni118, which was found to have highly temperature-sensitive coat protein but to cause green symptoms at the restrictive temperature (Jockusch, 1968). However, Ni118 did differ from the other isolates with temperature-sensitive coat proteins, in that it was by far the most temperature-sensitive of those tested, but did not differ from wild type in net electrical charge.

Wood and co-workers have studied the effects of mild and severe strains of CMV on various cellular functions. Ziemiecki and Wood (1976) could find no strain-related differences in radiolabelling of virus-stimulated proteins. Roberts and Wood (1981a) found that the severe strain tended to decrease synthesis of certain host proteins in the later stages of infection. They suggested that synthesis of excess coat protein by the severe strain might promote symptom formation. The milder strain caused increases in protease and ribonuclease activities, while the severe strain did not (Roberts and Wood, 1981b). Virus-induced changes in these degradative activities were therefore not positively correlated with symptom severity. Finally, Roberts and Wood (1982) showed that the chlorosis caused by the severe strain was associated with a reduction in the size of chloroplasts and fewer grana, but not with chloroplast disruption or accumulation of excess starch.

Many workers have studied changes in chlorophylls during infection and symptom development, and the results are discussed in connection with the effects of viruses on photosynthesis in Chapter 5.2.2. The effects of virus infections on other pigments have received less attention. Crosbie and Matthews (1974b) found that strains of TYMV causing pale green and yellow symptoms on Chinese cabbage leaves reduced the concentrations of chlorophylls a and b more than those of the four carotenoid pigments. A severe strain of TYMV causing white symptoms reduced the concentrations of all pigments equally.

7.3.4. Possible mechanisms controlling symptom formation at the molecular level

Van Telgen et al. (1985c) found evidence for an association between the coat proteins of TMV and two isolates of CMV causing different types of symptom, with the chromatin of infected tobacco plants. They suggested that this might be important in controlling symptom development, but did not elaborate on the possible mechanisms. Another report (Van Telgen et al., 1985a) suggested that the virus-coded 126 kDa polypeptide was also bound to the host chromatin, and that it too might control symptom formation.

The means by which viroids might induce symptoms have attracted much attention. Partly, this has been because of the small size of the pathogen and consequent ease of sequencing; mapping of the pathogenicity determinants of viroids is considered in the next section. However, the major stimulus to investigation of the problem must have been the intriguing challenge of explaining how these minimal pathogens can cause such extensive and devastating effects on the host. There have been various theories.

Since viroids and their (-) strand copies do not programme synthesis of any polypeptides (Sänger, 1982), their RNAs may cause symptoms directly, and the interaction with the host could involve either RNA sequence or secondary structure and thus molecular shape (Flores, 1984). It is likely, but not essential, that the effect on the host could involve interaction with host nucleic acids or a host nucleic acid-binding protein.

Schumacher *et al.* (1983) found that most of the PSTV in infected tomatoes occurred in the nucleoli, while Kiss *et al.* (1983) noted sequence homology between PSTV RNA and the small nuclear RNA (snRNA) U3B. Their U3B RNA sequence data were derived from Novikoff hepatoma (mammalian) cells. If the data can be taken as representative of a putative plant U3B-like RNA, they imply that viroids might interfere with its function. In mammalian cells, U3B is bound to the larger rRNA and its precursor in the nucleoli and may be involved in processing (reviewed in Busch *et al.*, 1982). Alternatively, Sänger (1984) has suggested that PSTV might complex with DNA-dependent RNA polymerase I (the nucleolar polymerase responsible for rRNA precursor synthesis) or its transcription factors, resulting in a competitive inhibition.

Several authors have suggested that viroids may be escaped regulatory RNAs, and that their pathogenicity is based on an aberrant regulatory function (Dickson, 1981; Diener, 1982). Viroids have been shown to have sequence homology with regions of mRNA precursors involved in splicing during removal of introns. The viroid (-) strand may also pair to the host U1a snRNA which is involved in mRNA processing (Mount *et al.*, 1983; Gross *et al.*, 1982; Gross, 1985; Solymosy and Kiss, 1985). A disruption of host mRNA processing would be highly likely to distort growth and development. A competitive model for inhibition of mRNA synthesis has also been proposed. Rackwitz *et al.* (1981) showed that host DNA-dependent RNA polymerase II, which synthesizes mRNA precursors, had a very high affinity for PSTV RNA, and suggested that this might inhibit mRNA synthesis *in vivo*.

Further aspects of possible interactions between viroids and host nucleic acids or proteins are considered in the next section in relation to mapping of the determinants of symptom type. It should also be noted that viroids have further, but probably indirect, effects on host metabolism which might contribute to symptom development. The pathogenesis-related proteins accumulated after viroid infection (discussed in Chapters 4.3.2; 4.5 and 6.6.2) cannot be specifically involved in the induction of symptoms by viroids, as they also accumulate in other types of infection. However, this does not exclude a possible role in development of symptoms in viroid infection. Also, plant growth regulator metabolism is altered in viroid-infected plants (Rodriguez et al., 1978; Duran-Villa and Semancik, 1982), and this could influence symptom development. The means by which viroids may alter protein synthesis and growth regulator metabolism are still unclear.

Low molecular weight RNAs (satellites) associated with viruses such as CMV (Waterworth et al., 1979) and TBSV (Hillman et al., 1985), and dependent on the virus for their replication, can alter the pattern of symptoms in the host. The symptoms can be made more severe or can become attenuated, and this may depend on the host involved. Thus CARNA-5, the satellite associated with CMV, can cause a lethal necrosis in tomato but attenuation of disease in other species (Waterworth et al., 1979). However, different isolates of the satellite caused different modifications of symptoms in a single host (Takanami, 1981; Kaper et al., 1981), and some satellites have no effect (Doz et al., 1980). Habili and Kaper (1981) found a correlation between the accumulation of the double-stranded form of CARNA-5 and attenuation of symptoms in tobacco. The modification of symptoms by CARNA-5 was strongly associated with reduction in both the mutliplication and specific infectivity of the helper virus (Piazolla et al., 1982). These authors also showed that CARNA-5 synthesis outcompeted CMV RNA synthesis. The molecular mechanisms by which satellites might modify development of helper virus symptoms are yet to be fully determined.

7.3.5. Mapping the determinants of symptom type on viroids and viruses

Viroids. Determination of the RNA sequences of a number of viroids or isolates, and calculations of secondary structure from predicted base–pairing, have indicated a remarkable level of similarity. Dickson et al. (1979) and Gross et al. (1981) showed that isolates of PSTV causing mild and severe symptoms differed in only a few nucleotides; the total sequence length of 359 was strongly conserved. Visvader and Symons (1985) examined 11 variants of CEV and found rather larger differences, including differences in sequence length. The sequences could be allocated into two groups which were also biologically distinct, producing severe and mild symptoms on tomato plants.

Fig. 7.4 shows the location of the nucleotide changes in isolates of PSTV causing symptoms of increasing severity; the effects on tomato hosts are shown in Fig. 7.5. It is clear that the sequence differences all mapped in the same region, and that increasing symptom severity was associated with a decrease in the calculated melting temperature of this region (Sänger, 1984). This region of the viroid has been termed the 'virulence modulating' (VM) region. Sequence analysis of other viroids has shown that almost all (nine of eleven compared) had sequences similar to the VM region of PSTV, in the same topological position on the calculated rod–shaped secondary structure (Schnölzer et al., 1985), although the other regions of the molecule could vary more. The two viroids which do not have VM–type sequences, ASBV and CCCV, showed very low overall sequence homology to the other nine and had completely different host ranges. They also had different intracellular locations and probably quite separate pathways of replication and symptom production (Schnölzer et al., 1985).

The apparent importance of conservation of the sequence of the VM region, and of secondary structure (Flores, 1984) in the group of nine viroids suggests that they are critical in determination of pathogenicity. From the correlation between the thermodynamic

Fig. 7.4. Location of nucleotide exchanges in various isolates of PSTV, causing mild, intermediate and severe symptoms on tomato, as shown in Fig. 7.5. The changes are relative to the mild isolate. The vertical lines indicate base-pairings, but the influence of the nucleotide exchanges on secondary structure is not shown. VM indicates the proposed 'virulence modulating' region. The calculated melting temperature of VM is shown. From Sänger (1984), by permission of the author and The Society for General Microbiology.

properties of the VM region and severity of symptoms produced by different isolates of PSTV (Sänger, 1984; Schnölzer et al., 1985), it was proposed that induction of disease symptoms might be related to the melting of the VM region in the plant. This implies that melting allowed association of the unpaired nucleotide sequence of the VM region with some host component, as yet unidentified. The extent of

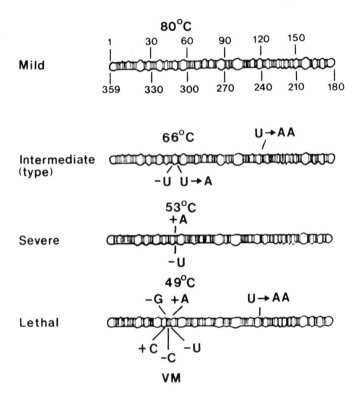

Lethal Intermediate Healthy
Severe Mild

PSTV isolate

Fig. 7.5. Tomato plants (cv. Rutgers) eight weeks after infection with PSTV isolates causing symptoms of varying severity. From Sänger (1984), by permission of the author and The Society for General Microbiology.

melting, and the nature of the sequence exposed, would then determine the pattern of symptom induction.

A problem with all these models of symptom induction by viroids is the range of types of symptom produced in the host by different isolates. Any model must be able to explain how very small changes in viroid sequence can control large changes in symptom type. This implies that the host recognition molecule must be very sensitive to small changes in the viroid. Another problem is to explain how the severity of symptoms caused by a viroid isolate is dependent on the host; different viroid isolates do not produce a constant ranking of symptom severities when grown in different hosts (Visvader and Symons, 1985).

Viruses. Most experiments have involved construction of

pseudorecombinants of viruses with multicomponent genomes, between isolates causing different symptoms, to localize the determinants of symptom type on particular genomic segments. Some of the work overlaps with the localization of determinants of virulence discussed in Chapter 6.5.2, especially where one of the symptoms is a local lesion resistance and the other systemic spread (e.g. Dingjan-Versteegh et al., 1972; De Jager and Van Kammen, 1970).

Other studies with pseudorecombinants have generally shown that symptoms can be controlled by determinants on more than one genomic segment, for example with TRV (Ghabrial and Lister, 1973), CLRV (Haber and Hamilton, 1980) and CPMV (De Jager and McLean, 1979). Habili and Francki (1974) constructed pseudorecombinants between CMV and the related virus TAV, and concluded that host reaction was determined by RNAs 1 and 2, whereas coat protein was determined by RNA 3. With other viruses, Bancroft and Lane (1973) showed that the RNA containing the coat protein gene could also determine symptoms. The clear message is that symptoms can be influenced by many different viral factors.

Sequencing of virus isolates giving different symptoms is beginning to give more precise information on the location of symptom determinants, and possibly on the viral function which is altered. The association of attenuation of systemic mosaic symptoms with a single base alteration in TMV RNA (Nishiguchi et al., 1985) was discussed in Chapter 6.6.4. Balazs et al. (1982) studied the nucleotide sequence of a CaMV isolate which caused a different pattern of symptoms, and compared the sequence with wild type. The two differed in about 5% of positions, and the differences were scattered throughout the genome, except that the two intergeneric regions were very highly conserved. Most of the sequence variation in the open reading frames was in the third position of codons, and would not have altered the coding.

7.3.6. Exogenous chemicals and environmental treatments which influence symptoms

There have been numerous experiments in which chemicals, especially those with activity as plant growth regulators, have been applied to plants in attempts to prevent or alter symptom development. Some of the effects were reviewed by Fraser and Whenham (1982). Generally, they have given little insight into how symptoms are caused in normal pathogenesis. Some workers have drawn parallels between virus-induced symptoms and the normal events of leaf senescence. ABA has been applied in an attempt to accelerate senescence (Balazs et al., 1973) and cytokinins in an attempt to delay it (Balazs and Kiraly, 1981). Some interactions have been found, but these have not been consistent, with opposing results being obtained with other combinations of hosts and viruses (e.g. Bailiss et al., 1977). It is also worth stressing that exogenous hormones may be taken up and metabolized differently from endogenous materials, and that experiments involving dose rates which are vastly different from natural concentrations are inherently unrealistic.

Environmental factors, such as temperature, light intensity and plant nutrition, can strongly influence symptom development and severity. An indication of the quantitative effects of different temperatures is shown in Fig. 6.8; the severity of systemic mosaic symptoms increased steeply with increasing growth temperature. Symptoms may also be altered in a qualitative manner, as in the change from a local-lesion response to a systemic infection with increasing temperature described in Chapter 6.4.2.

7.4. CONCLUSION

This has been, arguably, the most diverse Chapter in the whole book; there are so many different ways for viruses to influence host growth and development, and so many pathways to symptom formation. Some of these aspects are inherently difficult to study in hard molecular terms, as they involve changes which can, at present, only be expressed qualitatively or descriptively. There is therefore a need for more quantitative and mechanistic studies of the effects of viruses.

The second block to understanding how viruses control plants is that the relevant processes in the healthy host are not yet fully described in terms of molecular controls. This is most true of the hormonal control of growth and development: there is a need for more knowledge of how plant growth regulators work. Perhaps viruses will be useful in these studies.

CHAPTER 8

Viruses and Plant Biochemistry: Progress and Predictions

It should be obvious from the preceding Chapters that viruses affect diverse aspects of host metabolism, and induce both direct and secondary changes. The purpose of this final Chapter is briefly to highlight areas where more understanding is required, and to make some predictions about future developments.

Virus replication is still largely mysterious. Sequence homologies between the RNAs of plant viruses and regions coding for products with known functions in animal viruses, may prove one route to identification of virus-specified replicase components. Synthetic polypeptides, prepared to the amino acid sequence predicted by the nucleic acid base sequence, may be used to generate antibodies which will help in the isolation of replicases from infected plants. This may also provide an indirect means of isolating any host-coded components.

The broad details of the effects of virus infection on host nucleic acid metabolism have now been established, although for a limited number of hosts and viruses. The exciting challenges for the future are in elucidating the detailed effects of infection on the patterns of gene expression and activity, and the mechanisms involved. Recent studies of association of virus-coded proteins with chromatin are certainly worth confirming and should be expanded. Modern methods of analysis of genome activity, using cDNA probes for specific sequences, should give a wealth of information about individual genes.

At the protein level, there is an abundance of information on changes in metabolism of specific proteins or enzymes. What is needed are explanations of phenomena in the biological area, which probably involve protein-protein interactions. These include discovering how resistance mechanisms work, and the whole question of plant-virus recognition events in determination of host range, and resistance or susceptibility.

Progress in understanding control of growth and symptom development is patchy. Experiments with viroids are providing a fairly comprehensive explanation of how symptoms are caused, in molecular terms, although there are still gaps to be filled. Improved methodology is also giving insight into how virus-induced changes in plant growth regulators may affect growth and cause symptoms; one limit to progress here is that the metabolism and function of these compounds in healthy plants are incompletely understood. There are still deficiencies in our understanding of the links between viral effects on genome expression, and consequent development of symptoms.

Inevitably, most attention has centered on viruses as loss-causing agents, and their effects on host metabolism have been discussed largely in this context. Progress in understanding the molecular biology of virus replication may offer new ways to reduce crop losses. The most optimistic outlook is that 'molecular spanners' designed to interfere with virus multiplication may eventually lead to the extermination of some major virus diseases. There is also the possibility that viruses, or virus-derived systems, will find use in plant genetic engineering. The possibilities of using viruses as gene vectors may become less attractive, as other methods of DNA transfer are developed. However, parts of viral genomes are accessible and can be useful. The CaMV promoter has already been widely used, together with sequences from the Agrobacterium tumefaciens tumour-inducing plasmid, to regulate expression of bacterial and plant genes in plants. Other genes derived from viruses may prove useful in other contexts.

Finally, it is interesting to take a longer and less serious historical perspective. Mankind has doubtless had to endure crop

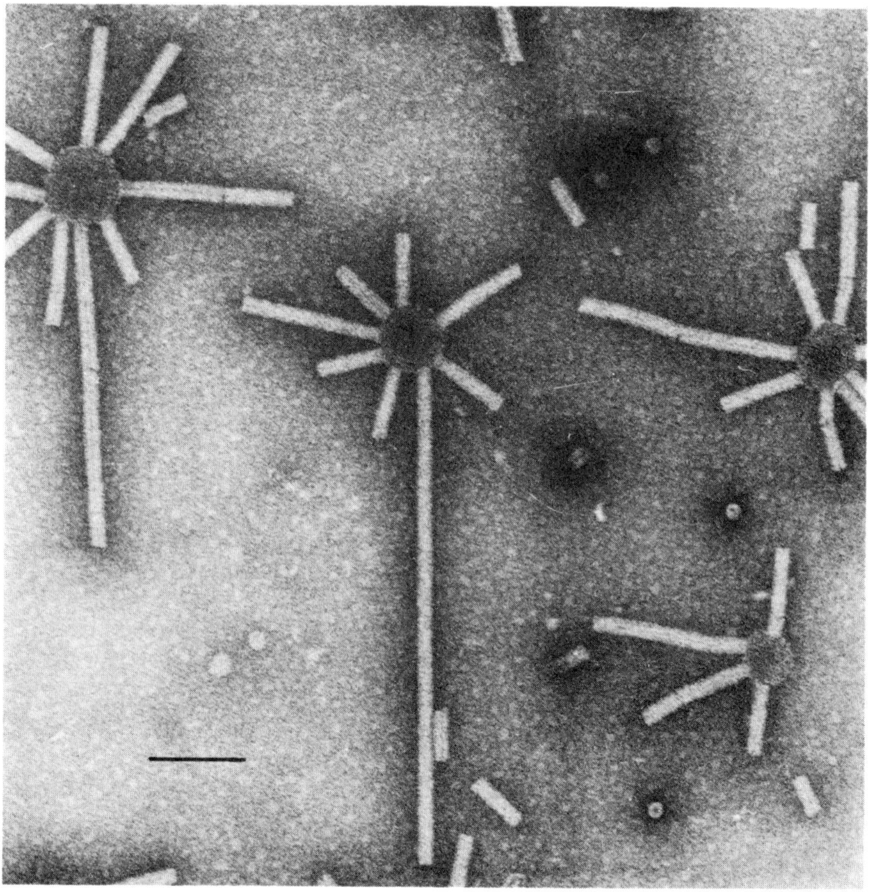

Fig. 8.1. Coat protein subunits of tobacco mosaic virus, assembled *in vitro* into helical rods. The rods are of variable length, but otherwise resemble virus particles in morphology. The assembled rods have adhered to latex beads used as size markers, to give unusual star-like clusters. The bar shows 100 nm.

losses from plant viruses since the beginnings of cultivation, but paradoxically, our first record is of a beneficial infection. In early 17th Century Holland, tulips with flower-break, a striping of petal colour caused by infection with tulip breaking virus, were highly regarded. Infected bulbs commanded very high prices, and infected flowers feature in several paintings from that time.

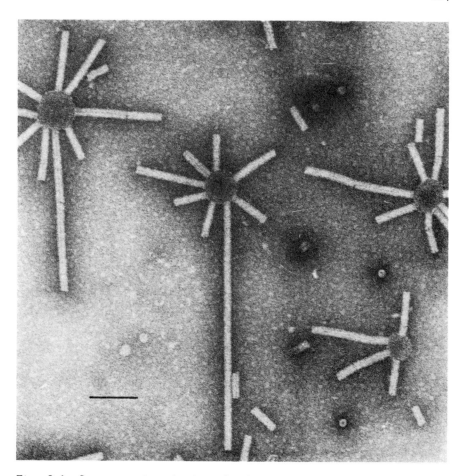

Modern molecular biology has given us an awesome understanding of plant virus structure and an ability to re-create some parts of it in the test tube. The reassembled TMV coat protein shown in Fig. 8.1 is interesting not only because of the scientific content, but because it also has a certain attraction to the eye. There is a pleasing symmetry in the aesthetic appeal of flower break and the later molecular construction, despite their separation in time and methodology. It is also notable that while the molecular biology of coat protein assembly is well understood, the biochemical mechanisms by which viruses cause flower breaking are still to be completely unravelled.

Bibliography

Abel, P.P., Nelson, R.S., De, B., Hoffmann, N., Rogers, S.G., Fraley, R.T., and Beachy, R.N. (1986). Science 232, 738–743.

Abu-Jawdah, Y. (1982). Changes in the soluble protein patterns of bean leaves upon fungal or viral infections or after chemical injury. Phytopathol. Z. 103, 272–279.

Adams, J.B., and Wade, C.V. (1976). Aphid behaviour and host plant preference demonstrated by electronic patterns of probing and feeding. Am. Potato J. 53, 261–267.

Agrios, G.N., Walker, M.E., and Ferro, D.N. (1985). Effect of cucumber mosaic virus inoculation at successive weekly intervals on growth and yield of pepper (Capsicum annuum) plants. Plant Dis. 69, 52–55.

Aharoni, N., Marco, S., and Levy, D. (1977). Involvement of gibberellins and abscisic acid in suppression of hypocotyl elongation in CMV-infected cucumbers. Physiol. Plant Pathol. 11, 189–194.

Ahl, P., Benjama, A., Samson, R., and Gianinazzi, S. (1981). New host proteins (b-proteins) induced together with resistance to a secondary infection following a bacterial infection in tobacco. Phytopathol. Z. 102, 201–212.

Ahl, P., Cornu, A., and Gianinazzi, S. (1982). Soluble proteins as genetic markers in studies of resistance and phylogeny in Nicotiana. Phytopathology 72, 80–85.

Ahl, P., and Gianinazzi, S. (1982). b-Proteins as a constitutive component in highly (TMV) resistant interspecific hybrids of N. glutinosa x N. debneyi. Pl. Sci. Lett. 26, 173–181.

Altman, A., Friedman, R., Amir, D., and Levin, N. (1982). Polyamine effects and metabolism in plants under stress conditions. In Plant Growth Substances 982. (Edited by P.F. Wareing). pp. 483–494. Academic Press, London.

Amemiya, Y., and Misawa, T. (1977). Studies on the resistance of cucumber to cucumber mosaic virus. II. Induction of resistance by infection. Tohoku J. Agric. Res. 28, 18–25.

Andebrhan, T., Coutts, R.H.A., Wagih, E.E., and Wood, R.K.S. (1980). Induced resistance and changes in the soluble protein fraction of cucumber leaves locally infected with Colletotrichum lagenarium or tobacco necrosis virus. Phytopathol. Z. 98, 47–52.

Anderson, J.M. (1982). The role of chlorophyll-protein complexes in the function and structure of chloroplast thylakoids. Mol. Cell Biochem. 46, 161–172.

Ansa, O.A., Bowyer, J.W., and Shepherd, R.J. (1982). Evidence for replication of cauliflower mosaic virus DNA in plant nuclei. Virology 121, 147–156.

Antignus, Y., Sela, I., and Harpaz, I. (1975). A phosphorus-containing fraction associated with antiviral activity in Nicotiana spp. carrying the gene for localization of TMV infection. Physiol. Plant Pathol. 6, 159-168.

Antignus, Y., Sela, I., and Harpaz, I. (1977). Further studies on the biology of an antiviral factor (AVF) from virus infected plants and its association with the N-gene of Nicotiana species. J. Gen. Virol. 35, 107-116.

Antoniw, J.F., Kueh, J.S.H., Walkey, D.G.A., and White, R.F. (1981). The presence of pathogenesis-related proteins in callus of Xanthi-nc tobacco. Phytopathol. Z. 101, 179-184.

Antoniw, J.F., Ritter, C.E., Pierpoint, W.S., and Van Loon, L.C. (1980). Comparison of three pathogenesis-related proteins from plants of two cultivars of tobacco infected with TMV. J. Gen. Virol. 47, 79-87.

Antoniw, J.F., and White, R.F. (1983). Biochemical properties of the pathogenesis-related proteins from tobacco. Neth. J. Plant Pathol. 89, 255-264.

Antoniw, J.F., White, R.F., and Carr, J.P. (1984). An examination of the effect of human γ-interferons on the infection and multiplication of tobacco mosaic virus in tobacco. Phytopathol. Z. 109, 367-371.

Appiano, A., and D'Agostino, G. (1985). Myelinic and granular inclusion bodies in non-inoculated tissues of both healthy and TMV-infected leaves of Xanthi n.c. tobacco. Neth. J. Plant Pathol. 91, 163-168.

Appiano, A., D'Agostino, G., Redolfi,. P., and Pennazio, S. (1981). Sequence of cytological events during the process of local lesion formation in the tomato bushy stunt virus-Gomphrena globosa hypersensitive system. J. Ultrastruct. Res. 76, 173-180.

Appiano, A., Pennazio, S., D'Agostino, G., and Redolfi, P. (1977). Fine structure of necrotic local lesions induced by tomato bushy stunt virus in Gomphrena globosa leaves. Physiol. Plant Pathol. 11, 327-332.

Appiano, A., Pennazio, S., and Redolfi, P. (1978). Cytological alterations in tissues of Gomphrena globosa plants systemically infected with tomato bushy stunt virus. J. Gen. Virol. 40, 277-286.

Astier-Manifacier, S., and Cornuet, P. (1971). RNA-dependent RNA polymerase in Chinese cabbage. Biochem. Biophys Acta 232, 484-493.

Atabekov, J.G. (1975). Host specificity of plant viruses. Annu. Rev. Phytopathol. 13, 127-145.

Atabekov, J.G., and Dorokhov, Y.L. (1984). Plant virus-specific transport function and resistance of plants to viruses. Adv. Virus Res. 29, 313-363.

Atabekov, J.G., Novikov, V.K., Vishnichenko, V.K., and Javakhia, V.G. (1970). A study of the mechanisms controlling the host range of plant viruses. II. The host range of hybrid viruses reconstituted in vitro and of free viral RNA. Virology 41, 108-115.

Atchison, B.A. (1973). Division, expansion and DNA synthesis in meristematic cells of French bean (Phaseolus vulgaris L.) root-tips invaded by tobacco ringspot virus. Physiol. Plant Pathol. 3, 1-8.

Atkinson, P.H., and Matthews, R.E.F. (1970). On the origin of dark green tissue in tobacco leaves infected with tobacco mosaic virus.

Virology 40, 344-356.
Babos, P. (1966). The ribonucleic acid content of tobacco leaves infected with tobacco mosaic virus. Virology 28, 282-289.
Bailiss, K.W. (1968). Gibberellins and the early disease syndrome of aspermy virus in tomato (Lycopersicon esculentum Mill.). Ann. Bot. 32, 543-552.
Bailiss, K.W. (1974). The relationship of gibberellin content to cucumber mosaic virus infection of cucumber. Physiol. Plant Pathol. 4, 73-79.
Bailiss, K.W. (1977). Gibberellins, abscisic acid and virus-induced stunting. In Current Topics in Plant Pathology. (Edited by Z. Kiraly). pp. 361-373. Academiai Kiado, Budapest.
Bailiss, K.W., Cocker, F.M., and Cassells, A.C. (1977). The effect of benlate and cytokinins on the content of tobacco mosaic virus in tomato leaf discs and cucumber mosaic virus in cucumber cotyledon discs and seedlings. Ann. Appl. Biol. 87, 383-392.
Bailiss, K.W., and Senananyake, S. (1984). Virus infection and reproductive losses in faba bean (Vicia faba L.). Plant Pathol. 33, 185-192.
Balazs, E., Barna, B., and Kiraly, Z. (1976). Effect of kinetin on lesion development and infection sites in Xanthi-nc tobacco infected with TMV: single cell local lesions. Acta Phytopathol. Acad. Sci. Hung. 11, 1-9.
Balazs, E., Barna, B., and Kiraly, Z. (1977). Heat-induced local lesions with high peroxidase activity in a systemic host of TMV. Acta Phytopathol. Acad. Sci. Hung. 12, 151-156.
Balazs, E., Gaborjanyi, R., and Kiraly, Z. (1973). Leaf senescence and increased virus susceptibility in tobacco: the effect of abscisic acid. Physiol. Plant Pathol. 3, 341-346.
Balazs, E., Gaborjanyi, R., and Tyihak, E. (1979). Histone-like proteins in tobacco plants infected with tobacco mosaic virus. Acta Phytopathol. Acad. Sci. Hung. 14, 239-246.
Balazs, E., Guilley, H., Jonard, G., and Richards, K. (1982). Nucleotide sequence of DNA from an altered-virulence isolate D/H of the cauliflower mosaic virus. Gene 19, 239-249.
Balazs, E., and Kiraly, Z. (1981). Virus content and symptom expression in Samsun tobacco treated with kinetin and a benzimidazole derivative. Phytopathol. Z. 100, 356-360.
Balazs, E., Sziraki, I., and Kiraly, Z. (1977). The role of cytokinins in the systemic acquired resistance of tobacco hypersensitive to tobacco mosaic virus. Physiol. Plant Pathol. 11, 29-37.
Bald, J.G., and Tinsley, T.W. (1967a). A quasi-genetic model for plant virus host ranges. I. Group reactions within taxonomic boundaries. Virology 31, 616-624.
Bald, J.G., and Tinsley, T.W. (1967b). A quasi-genetic model for plant virus host ranges. II. Differentiation between host ranges. Virology 32, 321-327.
Bald, J.G., and Tinsley, T.W. (1967c). A quasi-genetic model for plant virus host ranges. III. Congruence and relatedness. Virology 32, 328-336.
Balint, R., and Cohen, S.S. (1985). The incorporation of

radiolabelled polyamines and methionine into turnip yellow mosaic virus in protoplasts from infected plants. Virology 144, 181-193.

Bancroft, J.B., and Lane, L.C. (1973). Genetic analysis of cowpea chlorotic mottle and brome mosaic virus. J. Gen. Virol. 19, 381-389.

Bancroft, J.B., Motoyoshi, F., Watts, J.W., and Dawson, J.R.O. (1975). Cowpea chlorotic mottle and brome mosaic viruses in tobacco protoplasts. In Modification of the Information Content of Plant Cells. (Edited by R. Markham, D.R. Davies, D.A.Hopwood and R.W. Horne). pp. 133-160. Elsevier/North Holland, Amsterdam.

Banerjee, S., Vandenbranden, M., and Ruysschaert, J.M. (1981). Interaction of tobacco mosaic virus protein with lipid membrane systems. FEBS Lett. 133, 221-224.

Barbara, D.J., and Wood, K.R. (1974). The influence of actinomycin D on cucumber mosaic virus (strain W) multiplication in cucumber cultivars. Physiol. Plant Pathol. 4, 45-50.

Barker, H., and Harrison, B.D. (1978). Double infection, interference and superinfection in protoplasts exposed to two strains of raspberry ringspot virus. J. Gen. Virol. 40, 647-658.

Batra, G.K., and Kuhn, C.W. (1975). Polyphenoloxidase and peroxidase activities associated with acquired resistance and its inhibition by 2-thiouracil in virus-infected soybean. Physiol. Plant Pathol. 5, 239-248.

Baulcombe, D.C., Saunders, G.R., Bevan, M.W., Mayo, M.A., and Harrison, B.D. (1986). Expression of biologically active viral satellite RNA from the nuclear genome of transformed plants. Nature 321, 446-449.

Bedbrook, J.R., and Matthews, R.E.F. (1972). Changes in the proportions of early products of photosynthetic carbon fixation induced by TYMV infection. Virology 48, 255-258.

Bedbrook, J.R., and Matthews, R.E.F. (1973). Changes in the flow of early products of photosynthetic carbon fixation associated with replication of TYMV. Virology 53, 84-91.

Bennett, J., and Scott, K.J. (1971). Ribosome metabolism in mildew-infected barley leaves. FEBS Lett. 16, 93-95.

Benson, A.A., and Jekela, A.T. (1976). Cell membranes, in Plant Biochemistry (3rd Edition). (Edited by J. Bonner and J.E. Varner). pp. 65-89. Academic Press, New York.

Ben-Tal, Y., and Marco, S. (1980). Qualitative changes in cucumber gibberellins following cucumber mosaic virus infection. Physiol. Plant Pathol. 16, 327-336.

Bergman, E.L., and Boyle, J.S. (1962). Effect of tobacco mosaic virus on the mineral content of tomato leaves. Phytopathology 52, 956-957.

Bialy, H., and Klausner, A. (1986). A new route to virus resistance in plants. Bio/Technology 4, 96.

Biddington, N.L., and Thomas, T.H. (1978). Influence of different cytokinins on the transpiration and senescence of excised oat leaves. Physiol. Plant. 42, 369-374.

Bijaisoradat, M., and Kuhn, C.W. (1985). Nature of resistance in soybean to cowpea chlorotic mottle virus. Phytopathology 75, 351-355.

Boege, F., Rohde, W., and Sänger, H.L. (1982). In vitro transcription of viroid RNA into full length copies by RNA-dependent RNA polymerase from healthy tomato leaf tissue. Biosci. Rep. 2, 185-194.

Bol, J.F., Batshuizen, C.E.G.C., and Rutgers, T. (1976). Composition and biosynthetic activity of polyribosomes associated with alfalfa mosaic virus infections. Virology 75, 1-17.

Boninsegna, J.A., and Sayavedra, E. (1978). Starch metabolism in healthy and tomato bushy stunt virus-infected Lycopersicon esculentum plants. Phytopathol. Z. 91, 163-169.

Bos, L. (1978). Symptoms of Virus Diseases in Plants. Third Edition. 225 pp. Centre for Agricultural Publication and Documentation, Wageningen.

Boulton, M.I., Maule, A.J., and Wood, K.R. (1985). Effects of actinomycin D and UV irradiation on the replication of cucumber mosaic virus in protoplasts isolated from resistant and susceptible cucumber cultivars. Physiol. Plant Pathol. 26, 279-288.

Bozarth, R.F., and Diener, T.O. (1963). Changes in concentrations of free amino acids and amides induced in tobacco plants by potato virus X and potato virus Y. Virology 21, 188-193.

Brakke, M.K. (1984). Mutations, the abberrant ratio phenomenon, and virus infection of maize. Annu. Rev. Phytopathol. 22, 77-94.

Branch, A.D., and Robertson, H.D. (1984). A replication cycle for viroids and other small infectious RNAs. Science 223, 450-455.

Branch, A.D., Robertson, H.D., Greer, C., Gegenheimer, P., Peebles, C., and Abelson, J. (1982). Cell free circularization of viral progeny RNA by an RNA ligase from wheat germ. Science 217, 1147-1149.

Brisco, M.J., Hull, R., and Wilson, T.M.A. (1985). Southern bean mosaic virus-specific proteins are synthesized in an in vitro system supplemented with intact, treated virions. Virology 143, 392-398.

Brishammar, S. (1976). Separation studies of TMV replicase. Ann. Microbiol. 127A, 25-31.

Brisson, L., Asselin, A., and Trudel, M.J. (1984). Influence of tomato mosaic virus on the yield of 4 cultivars of Lycopersicon esculentum. Phytoprotection 65, 74-80.

Broadbent, L. (1964). The epidemiology of tomato mosaic. VII. The effect of TMV on tomato fruit yield and quality under glass. Ann. Appl. Biol. 54, 209-224.

Brown, E.G., and Newton, R.P. (1981). Cyclic AMP and higher plants. Phytochemistry 20, 2453-2463.

Brown, F., and Martin, S.J. (1965). A new model for virus ribonucleic acid replication. Nature 208, 861-863.

Bujarski, J.J., Hardy, S.F., Miller, W.A., and Hall, T.C. (1982). Use of dodecyl-β-D-maltoside in the purification and stabilization of RNA polymerase from brome mosaic virus-infected barley. Virology 119, 465-473.

Bukrinskaya, A.G. (1982). Penetration of viral genetic material into host cell. Adv. Virus Res. 27, 141-204.

Busch, H., Raddy, R., Rothblum, L., and Choi, Y.C. (1982). SnRNAs,

SnRNPs, and RNA processing. Annu. Rev. Biochem. 51, 617-654.
Butler, P.J.G. (1984). The current picture of the assembly of tobacco mosaic virus. J. Gen. Virol. 65, 253-279.
Butler, W.J. (1978). Energy distribution in the photochemical apparatus of photosynthesis. Annu. Rev. Plant Physiol. 29, 345-378.
Camacho Henriquez, A.C., Lucas, J., and Sänger, H.L. (1983). Purification and biochemical properties of the 'pathogenesis-related' protein p14 from tomato leaves. Neth. J. Plant Pathol. 89, 308.
Camacho Henriquez, A., and Sänger, H.-L. (1982a). Gel electrophoretic analysis of phenol-extractable leaf proteins from different viroid-host combinations. Arch. Virol. 74, 167-180.
Camacho Henriquez, A., and Sänger, H.-L. (1982b). Analysis of acid-extractable tomato leaf proteins after infection with a viroid, two viruses and a fungus and partial purification of the pathogenesis-related protein p14. Arch. Virol. 74, 181-196.
Camacho Henriquez, A.C., and Sänger, H.L. (1984). Purification and partial characterization of the major "pathogenesis-related" tomato leaf protein P14 from potato spindle tuber viroid (PSTV)-infected tomato leaves. Arch. Virol. 81, 263-284.
Carlson, P.S., and Murakishi, H.H. (1978). Evidence on the clonal versus non-clonal origin of dark green islands in virus-infected tobacco leaves. Plant Sci. Lett. 13, 377-381.
Carr, J.P. (1985). The control of synthesis of the pathogenesis-related proteins of tobacco. Ph.D. Thesis, University of Liverpool, U.K.
Carr, J.P., Antoniw, J.F., White, R.F., and Wilson, T.M.A. (1982). Latent messenger RNA in tobacco (Nicotiana tabacum). Biochem. Soc. Trans. 10, 353-354.
Carroll, T.W. (1970). Relation of barley stripe mosaic virus to plastids. Virology 42, 1015-1022.
Carroll, T.W., Gossel, P.L., and Hockett, E.A. (1979). Inheritance of resistance to seed transmission of barley stripe mosaic virus in barley. Phytopathology 69, 431-433.
Cassells, A.C. (1983). Chemical control of virus disease of plants. Prog. Med. Chem. 20, 119-155.
Cassells, A.C., Barnett, A., and Barlass, M. (1978). The effect of polyacrylic acid treatment on the susceptibility of Nicotiana tabacum cv. Xanthi-nc to tobacco mosaic virus. Physiol. Plant Pathol. 13, 13-21.
Cassells, A.C., and Herrick, C.C. (1977a). Cross protection between mild and severe strains of tobacco mosaic virus on doubly inoculated tomato plants. Virology 78, 253-260.
Cassells, A.C., and Herrick, C.C. (1977b). The identification of mild and severe strains of tobacco mosaic virus in doubly inoculated tomato plants. Ann. Appl. Biol. 86, 37-46.
Cassells, A.C., and Minas, G. (1983). Beneficially infected and chimeral Pelargonium: implications for micropropagation by meristem and explant culture. Acta Hortic. 131, 287-298.
Caylay, P.J., White, R.F., Antoniw, J.F., Walesby, N.J., and Kerr, I.M. (1982). Distribution of the ppp(A2'p5')$_n$-binding protein and

interferon-related enzymes in animals, plants and lower organisms. Biochem. Biophys. Res. Commun. 108, 1243-1250.

Chadha, K.C., and MacNeill, B.H. (1969). An antiviral principle from tomatoes systemically infected with tobacco mosaic virus. Ind. J. Bot. 47, 513-518.

Chalcroft, J.P., and Matthews, R.E.F. (1967a). Virus strain and leaf ontogeny as factors in the production of leaf mosaic patterns by turnip yellow mosaic virus. Virology 33, 167-171.

Chalcroft, J.P., and Matthews, R.E.F. (1967b). Role of virus strain and leaf ontogeny in the production of mosaic patterns by turnip yellow mosaic virus. Virology 33, 659-673.

Chandra, S., and Mondy, N.I. (1981). Effect of potato virus X on the mineral content of potato tubers. J. Agric. Food Chem. 29, 811-814.

Chant, S.R. (1983). Effect of nutrition on the susceptibility of cowpeas to virus infection and multiplication. Microbiol. Lett. 93, 39-48.

Chessin, M. (1982). Interference in plant virus infection: ultraviolet light and systemic acquired resistance. Phytopathol. Z. 104, 279-283.

Chessin, M. (1983). Is there a plant interferon? Bot. Rev. 49, 1-27.

Chifflot, S., Sommer, P., Hartmann, D., Stussi-Garaud, C., and Hirth, L. (1980). Replication of alfalfa mosaic virus RNA: evidence for a soluble replicase in healthy and infected tobacco leaves. Virology 100, 91-100.

Cogdell, R.J. (1983). Photosynthetic reaction centers. Annu. Rev. Plant Physiol. 34, 21-45.

Cohen, S.S., and Greenberg, M.L. (1981). Spermidine, an intrinsic component of turnip yellow mosaic virus. Proc. Natl. Acad. Sci. USA 78, 5470-5474.

Coleman, J., Hirashima, A., Inokuchi, Y., Green, P.J., and Inouye, M. (1985). A novel immune system against bacteriophage infection using complementary RNA (micRNA). Nature 315, 601-603.

Collendavelloo, J., Legrand, M., and Fritig, B. (1983). Plant disease and the regulation of enzymes involved in lignification. Increased rate of de novo synthesis of the three tobacco O-methyltransferases during the hypersensitive response to infection by tobacco mosaic virus. Plant Physiol. 73, 550-554.

Commoner, B., and Nehari, V. (1953). The effects of tobacco mosaic virus synthesis on the free amino acid and amide composition of the host. J. Gen. Physiol. 36, 791-805.

Conejero, V., Picazo, I., and Segado, P. (1979). Citrus exocortis viroid (CEV): protein alterations in different hosts following viroid infection. Virology 97, 454-456.

Conejero, V., and Semancik, J.S. (1977). Exocortis viroid: alteration in the proteins of _Gynura aurantiaca_ accompanying viroid infection. Virology 77, 221-232.

Cooke, R., Durand, R., Teissere, M., Penon, P., and Ricard, J. (1981). Characterization of heparin-resistant complex formation and RNA synthesis by wheat germ RNA polymerases I, II and III, _in vitro_ on cauliflower mosaic virus DNA. Biochem. Biophys. Res. Commun. 98, 36-42.

Cooper, J.I., and Jones, A.T. (1983). Responses of plants to viruses: proposals for the use of terms. Phytopathology 73, 127-128.

Costa, A.S., and Müller, G.W. (1980). Tristeza control by cross protection: a U.S.-Brazil cooperative success. Plant Dis. 64, 538-541.

Couch, H.B. (1955). Studies on seed transmission of lettuce mosaic virus. Phytopathology 45, 63-71.

Coutts, R.H.A., and Wagih, E.E. (1983). Induced resistance to viral infection and soluble protein alterations in cucumber and cowpea plants. Phytopathol. Z. 107, 57-69.

Coutts, R.H.A., and Wood, K.R. (1977). Inoculation of leaf mesophyll protoplasts from a resistant and a susceptible cucumber cultivar with cucumber mosaic virus. FEMS Microbiol. Lett. 1, 121-124.

Covey, S.N., and Hull, R. (1981). Transcription of cauliflower mosaic virus DNA. Detection of transcripts, properties, and location of the gene encoding the virus inclusion body protein. Virology 111, 463-474.

Crosbie, E.S., and Matthews, R.E.F. (1974a). Effects of TYMV infection on growth of Brassica pekinensis Rupr. Physiol. Plant Pathol. 4, 389-400.

Crosbie, E.S., and Matthews, R.E.F. (1974b). Effects of TYMV infection on leaf pigments in Brassica pekinensis Rupr. Physiol. Plant Pathol. 4, 379-387.

Cruikshank, I.A.M., and Perrin, D.R. (1964). Pathological functions of phenolic compounds in plants, in Biochemistry of Phenolic Compounds. (Edited by J.B. Harborne). pp. 511-544. Academic Press, New York.

Da Graca, J.V., and Martin, M.M. (1976). An electron microscope study of hypersensitive tobacco infected with tobacco mosaic virus at 32°C. Physiol. Plant Pathol. 8, 215-219.

Da Graca, J.V., and Martin, M.M. (1981). Ultrastructural changes in avocado leaf tissue infected with avocado sunblotch. Phytopathol. Z. 102, 185-194.

Damirdagh, I.S., and Ross, A.F. (1967). A marked synergistic reaction of potato viruses X and Y in inoculated leaves of tobacco. Virology 31, 296-307.

Darby, L.A., Ritchie, D.B., and Taylor, I.B. (1978). Isogenic lines of the tomato 'Ailsa Craig'. Rep. Glasshouse Crops Res. Inst. 1977, 168-184.

Dasgupta, A., Zabel, P., and Baltimore, D. (1980). Dependence of the activity of poliovirus replicase on a host cell protein. Cell 19, 423-429.

Davies, J.W. (Ed.) (1985a). Molecular Plant Virology, Vol. I. Virus Structure and Assembly and Nucleic Acid-Protein Interactions. CRC Press, Boca Raton. 240 pp.

Davies, J.W. (Ed.) (1985b). Molecular Plant Virology Vol. II. CRC Press, Boca Raton. 272 pp.

Davies, J.W., and Hull, R. (1982). Genome expression of plant positive-strand RNA viruses. J. Gen. Virol. 61, 1-14.

Davies, J.W., Stanley, J., and Van Kammen, A. (1979). Sequence homology adjacent to the 3' terminal poly(A) of cowpea mosaic virus RNAs. Nucleic Acids Res. 7, 493-500.

Davies, J.W., and Verduin, B.J.M. (1979). In vitro synthesis of cowpea chlorotic mottle virus polypeptides. J. Gen. Virol. 44, 545-549.

Dawkins, R. (1976). The Selfish Gene. Oxford University Press. 224 pp.

Dawson, G.W., Gibson, R.W., Griffiths, D.C., Pickett, J.A., Rice, A.D., and Woodcock, C.M. (1982). Aphid alarm pheromone derivatives affecting settling and transmission of some plant viruses. J. Chem. Ecol. 8, 1377-1387.

Dawson, J.R.O. (1965). Contrasting effects of resistant and susceptible tomato plants on tomato mosaic virus multiplication. Ann. Appl. Biol. 56, 485-491.

Dawson, J.R.O., Rees, M.W., and Short, M.N. (1975). Protein composition of unusual tobacco mosaic virus strains. Ann. Appl. Biol. 79, 189-194.

Dawson, J.R.O., Rees, M.W., and Short, M.N. (1979). Lack of correlation between the coat protein composition of tobacco mosaic virus isolates and their ability to infect resistant tomato plants. Ann. Appl. Biol. 91, 353-358.

Dawson, W.O., and Dodds, J.A. (1982). Characterization of sub-genomic double-stranded RNAs from virus-infected plants. Biochem. Biophys. Res. Commun. 107, 1230-1235.

Dawson, W.O., and Schlegel, D.E. (1976). The sequence of inhibition of tobacco mosaic virus synthesis by actinomycin-D, 2-thiouracil and cycloheximide in a synchronous infection. Phytopathology 66, 177-181.

Day, K.L. (1984). Resistance to bean common mosaic virus in Phaseolus vulgaris L. Ph.D. Thesis, University of Birmingham, U.K.

De Fazio, G. (1981). Cytokinin levels in healthy and bean golden mosaic virus (BGMV) infected bean plants (Phaseolus vulgaris L.). Rev. Bras. Bot. 4, 57-71.

De Jager, C.P., and McLean, L. (1979). Further genetic analysis of a temperature-sensitive mutant of cowpea mosaic virus. Virology 99, 167-169.

De Jager, C.P., and Van Kammen, A. (1970). The relationship between the components of cowpea mosaic virus. III. Location of genetic information for two biological functions in the middle component of CPMV. Virology 41, 281-287.

De Jager, C.P., and Wesseling, J.B.M. (1981). Spontaneous mutations in cowpea mosaic virus overcoming resistance due to hypersensitivity in cowpea. Physiol. Plant Pathol. 19, 347-358.

Dekhuijzen, H.M. (1976). Endogenous cytokinins in healthy and diseased plants. In Physiological Plant Pathology. (Edited by R. Heitefuss and P.H. Williams). pp. 526-559. Springer-Verlag, Berlin.

De Laat, A.M.M., and Van Loon, L.C. (1982). Regulation of ethylene biosynthesis in virus-infected tobacco leaves. II. Time course of levels of intermediates and in vivo conversion rates. Plant Physiol. 69, 240-245.

De Laat, A.M.M., and Van Loon, L.C. (1983). The relationship between stimulated ethylene production and symptom expression in virus-infected tobacco leaves. Physiol. Plant Pathol. 22, 261-273.

De Laat, A.M.M., Van Loon, L.C., and Vonk, C.R. (1981). Regulation of ethylene biosynthesis in virus-infected tobacco leaves. I. Determination of the role of methionine as the precursor of ethylene. Plant Physiol. 68, 256-261.

Dellaporta, S.L., Wood, J., Hicks, J.B., Mottinger, J.P., and Chomet, P.S. (1983). Molecular characterization of shrunken mutations in Zea mays associated with virus infection. In Plant Infectious Agents: Viruses, Viroids, Virusoids and Satellites. (Edited by H.D. Robertson, S.H. Howell, M. Zaitlin and R.L.Malmberg). pp. 130-135. Cold Spring Harbor Laboratory, New York.

De Maeyer, E., De Maeyer-Guignard, J., and Montagnier, L. (1971). Double-stranded RNA from rat liver induces interferon in rat cells. Nature New Biol. 229, 109-110.

De Wit, P.J.G.M., and Bakker, J. (1980). Differential changes in soluble tomato leaf proteins after inoculation with virulent and avirulent races of Cladosporium fulvum (syn. Fulva fulva). Physiol. Plant Pathol. 17, 121-130.

De Varennes, A., Davies, J.W., Shaw, J.G., and Maule, A.J. (1985). A reappraisal of the effect of actinomycin D and cordycepin on the multiplication of cowpea mosaic virus in cowpea protoplasts. J. Gen. Virol. 66, 817-825.

Devash, Y., Biggs, S., and Sela, I. (1982). Multiplication of tobacco mosaic virus in tobacco leaf discs is inhibited by (2'-5') oligoadenylate. Science 216, 1415-1416.

Devash, Y., Haushner, A., Sela, I., and Chakraburtty, K. (1981). The antiviral factor (AVF) from virus-infected plants induces discharge of histidinyl-TMV-RNA. Virology 111, 103-112.

De Zoeten, G.A., and Gaard, G. (1983). Mechanisms underlying systemic invasion of pea plants by pea enation mosaic virus. Intervirology 19, 85-94.

De Zoeten, G.A., and Fulton, R.W. (1975). Understanding generates possibilities. Phytopathology 65, 221-222.

De Zoeten, G.A., and Rettig, N. (1972). Plant and aphid protein patterns as influenced by pea enation mosaic virus. Phytopathology 62, 1018-1023.

Dickson, E. (1981). A model for the involvement of viroids in RNA splicing. Virology 115, 216-221.

Dickson, E., Robertson, H.D., Niblett, L.L., Horst, R.K., and Zaitlin, M. (1979). Minor differences between nucleotide sequences of mild and severe strains of potato spindle tuber viroid. Nature 277, 60-62.

Diener, T.O. (1981). Are viroids escaped introns? Proc. Natl. Acad. Sci. USA. 78, 5014-5015.

Diener, T.O. (1982). Viroids and their interactions with host cells. Annu. Rev. Microbiol. 36, 239-258.

Dingjan-Versteegh, A., Van Vloten-Doting, L., and Jaspers, E.M.J. (1972). Alfalfa mosaic virus hybrids constructed by exchanging nucleic acid components. Virology 49, 716-722.

Dodds, J.A. (1982). Cross-protection and interference between electrophoretically distinct strains of cucumber mosaic virus in tomato. Virology 118, 235-240.

Dodds, J.A., and Hamilton, R.I. (1972). The influence of barley stripe mosaic virus on the replication of tobacco mosaic virus in Hordeum vulgare L. Virology 50, 404-411.

Dodds, J.A., and Hamilton, R.I. (1974). Masking of the RNA genome of tobacco mosaic virus by the protein of barley stripe mosaic virus in doubly infected barley. Virology 59, 418-426.

Doke, N. (1983a). Generation of superoxide anion by potato tuber protoplasts during the hypersensitive response to hyphal wall components of Phytophthora infestans and specific inhibition of the reaction by suppressors of hypersensitivity. Physiol. Plant Pathol. 23, 359-367.

Doke, N. (1983b). Involvement of superoxide anion generation in the hypersensitive response of potato tuber tissues to infection with an incompatible race of Phytophthora infestans and to the hyphal wall components. Physiol. Plant Pathol. 23, 345-357.

Doke, N. & Hirai, T. (1970). Effects of tobacco mosaic virus infection on photosynthetic CO_2 fixation and $^{14}CO_2$ incorporation into protein in tobacco leaves. Virology 42, 68-77.

Dorssers, L., Van der Meer, J., Van Kammen, A., and Zabel, P. (1983a). The cowpea mosaic virus RNA replication complex and the host-encoded RNA-dependent RNA polymerase template complex are functionally different. Virology 125, 155-174.

Dorssers, L., Zabel, P., Van der Meer, J., and Van Kammen, A. (1983b). Isolation and characterization of the cowpea mosaic virus RNA replication complex. In Plant Infectious Agents: Viruses, Viroids, Virusoids and Satellites. (Edited by H.D. Robertson, S.H. Howell, M. Zaitlin and R.L. Malmberg). pp. 120-125. Cold Spring Harbor Laboratory, New York.

Dorssers, L., Van der Krol, S., Van der Meer, J., and Van Kammen, A. (1984). Purification of cowpea mosaic virus RNA replication complex: Identification of a virus-encoded 110,000-dalton polypeptide responsible for RNA chain elongation. Proc. Natl. Acad. Sci. USA. 81, 1951-1955.

Dougherty, W.G., and Hiebert, E. (1980). Translation of potyvirus RNA in a rabbit reticulocyte lysate: identification of nuclear inclusion proteins as products of tobacco etch virus RNA translation and cylindrical inclusion protein as a product of the potyvirus genome. Virology 104, 174-182.

Doz, B., Macquaire, G., Delbos, R., and Dunez, J. (1980). Characteristics and role of the RNA 3, a satellite RNA of tomato black ring virus. Ann. Virol. 131, 489-499.

Drijfhout, E. (1978). Genetic interaction between Phaseolus vulgaris and bean common mosaic virus with implications for strain identification and breeding for resistance. Agric. Res. Rep. (Wageningen) 872. 98 pp.

Duchesne, M., Fritig, B., and Hirth, L. (1977). Phenylalanine ammonia-lyase in tobacco mosaic virus-infected hypersensitive tobacco. Density labelling evidence of de novo synthesis. Biochim. Biophys. Acta. 485, 465-481.

Duda, C.T. (1976). Plant RNA polymerases. Annu. Rev. Plant Physiol. 27, 119-132.

Duran-Vila, N., and Semancik, J.S. (1982). Effects of exogenous auxins on tomato tissue infected with citrus exocortis viroid.

Phytopathology 72, 777-781.

Durham, A.C.H., and Hendry, D.A. (1977). Cation binding by tobacco mosaic virus. Virology 77, 510-519.

Dyson, J.G., and Chessin, M. (1961). Effect of auxins on virus-induced leaf abscission. Phytopathology 51, 195.

Edwards, M.C., Gonsalves, D., and Provvidenti, R. (1983). Genetic analysis of cucumber mosaic virus in relation to host resistance: location of determinants for pathogenicity to certain legumes and Lactuca saligna. Phytopathology 73, 269-273.

Ekpo, E.J.A., and Saettler, A.W. (1975). Multiplication and distribution of bean common mosaic virus in Phaseolus vulgaris. Plant Dis. Rep. 59, 939-943.

Ellis, R.J. (1981a). Inhibitors for studying chloroplast transcription and translation in vivo. In Methods in Chloroplast Molecular Biology. (Edited by M. Edelman, R.B. Hallick, and N.-H. Chua). pp. 559-564. Elsevier/North Holland, Amsterdam.

Ellis, R.J. (1981b). Chloroplast proteins: synthesis, transport and assembly. Annu. Rev. Plant Physiol. 32, 111-137.

El-Meleigi, M.A., Sheen, S.J., and Lowe, R.H. (1981). Soluble proteins and enzymes in the sub-cellular fractions of virus-infected tobacco leaf. Can. J. Plant Sci., 61, 135-142.

Endo, R.M., and Brown, C.M. (1963). Effects of barley yellow dwarf virus on yield of oats as influenced by variety, virus strain, and developmental stage of plants at inoculation. Phytopathology 53, 965-968.

Endo, T. (1968). Indoleacetic acid oxidase activity of horseradish and other plant peroxidase isoenzymes. Plant Cell Physiol. 9, 333-341.

Evans, D.M.A., Bryant, J.A., and Fraser, R.S.S. (1984). Characterization of RNA-dependent RNA polymerases in healthy and tobacco mosaic virus-infected plants. Ann. Bot. 54, 271-281.

Evans, D.M.A., Fraser, R.S.S., and Bryant, J.A. (1985). RNA-dependent RNA polymerase activities in tomato plants susceptible or resistant to tobacco mosaic virus. Ann. Bot. 55, 587-591.

Faccioli, G. (1979). Relationship of peroxidase, catalase and polyphenoloxidase to acquired resistance in plants of Chenopodium amaranticolor. Phytopathol. Z. 95, 237-249.

Fantes, K.H., and O'Neill, G.F. (1964). Some similarities between a virus inhibitor of plant origin and chick interferon. Nature 203, 1048-1050.

Farkas, G.L., and Stahmann, M.A. (1966). On the nature of changes in peroxidase isoenzymes in bean leaves infected by southern bean mosaic virus. Phytopathology 56, 669-677.

Faulkner, G., and Kimmins, W.C. (1975). Staining reactions to the tissue bordering virus lesions induced by wounding, tobacco mosaic virus and tobacco necrosis virus in bean. Phytopathology 65, 1396-1400.

Favali, M.A., Conti, G.G., and Bassi. M. (1978). Modifications of the vascular bundle ultrastructure in the "resistance zone" around necrotic lesions induced by tobacco mosaic virus. Physiol. Plant Pathol. 13, 247-251.

Federation of British Plant Pathologists. (1973). A guide to the use of terms in plant pathology. Phytopathol. Paper 17, Commonwealth Mycological Institute, Kew, Surrey, England.

Fernandez, T.F., and Gaborjanyi, R. (1976). Reversion of dwarfing induced by virus infection: effect of polyacrylic and gibberellic acids. Acta Phytopathol. Acad. Sci. Hung. 11, 271-275.

Fernandez-Gonzalez, O., Renaudin, J., and Bove, J. (1980). Infection of chlorophyll-less protoplasts from etiolated Chinese cabbage hypocotyls by turnip yellow mosaic virus. Virology 104, 262-265.

Flor, H.H. (1956). The complementary genetic systems in flax and flax rust. Adv. Genet. 8, 29-54.

Flores, R. (1984). Is the conformation of viroids involved in their pathogenicity? J. Theor. Biol. 108, 519-527.

Flores, R., and Rodriguez, J.L. (1981). Altered pattern of root formation on cuttings of Gynura aurantiaca infected by citrus exocortis viroid. Phytopathology 71, 964-966.

Flores, R., and Semancik, J. (1982). Properties of a cell-free system for synthesis of citrus exocortis viroid. Proc. Natl. Acad. Sci. USA. 79, 6285-6288.

Foa-Tomasi, L., Campadelli-Fiume, G., Barbieri, L., and Stirpe, F. (1982). Effect of ribosome-inactivating proteins on virus-infected cells. Inhibition of virus multiplication and of protein synthesis. Arch. Virol. 71, 323-332.

Francki, R.I.B. (1981) Plant virus taxonomy. In Handbook of Plant Virus Infections: Comparative Diagnosis. (Edited by E. Kurstak). pp. 3-16. Elsevier/North Holland, Amsterdam.

Francki, R.I.B. (Ed.) (1985). The Plant Viruses. Vol. I. Polyhedral Virions with Tripartite Genomes. Plenum Press, New York. 297 pp.

Francki, R.I.B., Kitajima, E.W., and Peters, D. (1981). Rhabdoviruses. In Plant Virus Infections: Comparative Diagnosis. (Edited by E. Kurstak). pp. 455-489. Elsevier/North Holland, Amsterdam.

Francki, R.I.B., Milne, R.G., and Hatta, T. (1985a). Atlas of Plant Viruses Vol. I. CRC Press, Boca Raton. 240 pp.

Francki, R.I.B., Milne, R.G., and Hatta, T. (1985b). Atlas of Plant Viruses Vol. II. CRC Press, Boca Raton. 304 pp.

Francki, R.I.B., and Randles, J.W. (1972). RNA-dependent RNA polymerase associated with particles of lettuce necrotic yellows virus. Virology 47, 270-275.

Francki, R.I.B., and Randles, J.W. (1973). Some properties of lettuce necrotic yellows virus RNA and its in vitro transcription by virion-associated transcriptase. Virology 54, 359-368.

Franssen, H., Goldbach, R., and Van Kammen, A. (1984a). Translation of bottom component RNA of cowpea mosaic virus in reticulocyte lysate: faithful proteolytic processing of the primary translation product. Virus Res. 1, 39-49.

Franssen, H., Lennissen, J., Goldbach, R., Lomonossoff, G., and Zimmern, D. (1984b). Homologous sequences in non-structural proteins from cowpea mosaic virus and picornavirus. EMBO J. 3, 855-861.

Fraser, L., and Matthews, R.E.F. (1979). Strain-specific pathways of cytological change in individual Chinese cabbage protoplasts infected with turnip yellow mosaic virus. J. Gen. Virol. 45,

Fraser, L., and Matthews, R.E.F. (1981). A rapid transient inhibition of leaf initiation induced by turnip yellow mosaic virus infection. Physiol. Plant Pathol. 19, 325-336.

Fraser. L.G., and Matthews, R.E.F. (1983). A rapid transient inhibition of leaf initiation by abscisic acid. Plant Sci. Lett. 29, 67-72.

Fraser, R.S.S. (1969). Effects of two TMV strains on the synthesis and stability of chloroplast ribosomal RNA in tobacco leaves. Mol. Gen. Genet. 106, 73-79.

Fraser, R.S.S. (1972). Effects of two strains of tobacco mosaic virus on growth and RNA content of tobacco leaves. Virology 47, 261-269.

Fraser, R.S.S. (1973a). The synthesis of tobacco mosaic virus RNA and ribosomal RNA in tobacco leaves. J. Gen. Virol. 18, 267-279.

Fraser, R.S.S. (1973b). Tobacco mosaic virus is not methylated and does not contain a polyadenylic acid sequence. Virology 56, 379-382.

Fraser, R.S.S. (1975). Studies on messenger and ribosomal RNA synthesis in plant tissue cultures induced to undergo synchronous cell division. Eur. J. Biochem. 50, 529-537.

Fraser, R.S.S. (1979). Systemic consequences of the local lesion reaction to tobacco mosaic virus in a tobacco variety lacking the N gene for hypersensitivity. Physiol. Plant Pathol. 14, 383-394.

Fraser, R.S.S. (1981). Evidence for the occurrence of the 'pathogenesis-related' proteins in leaves of healthy tobacco plants during flowering. Physiol. Plant Pathol. 19, 69-76.

Fraser, R.S.S. (1982). Are 'pathogenesis-related' proteins involved in acquired systemic resistance of tobacco plants to tobacco mosaic virus? J. Gen. Virol. 58, 305-313.

Fraser, R.S.S. (1983). Varying effectiveness of the N' gene for resistance to tobacco mosaic virus in tobacco infected with virus strains differing in coat protein properties. Physiol. Plant Pathol. 22, 109-119.

Fraser, R.S.S. (1985a). Mechanisms of induced resistance to virus disease. In Mechanisms of Resistance to Plant Diseases. (Edited by R.S.S. Fraser). pp. 373-404. Martinus Nijhoff/Dr W. Junk, Dordrecht.

Fraser, R.S.S. (1985b). Genetics of host resistance to viruses and of virulence. In Mechanisms of Resistance to Plant Diseases. (Edited by R.S.S. Fraser). pp. 62-79. Martinus Nijhoff/Dr W. Junk, Dordrecht.

Fraser, R.S.S. (1985c). Host range control and non-host immunity to viruses. In Mechanisms of Resistance to Plant Diseases. (Edited by R.S.S. Fraser). pp. 13-28. Martinus Nijhoff/Dr W. Junk, Dordrecht.

Fraser, R.S.S. (1985d). Mechanisms involved in genetically controlled resistance and virulence: virus diseases. In Mechanisms of Resistance to Plant Diseases. (Edited by R.S.S. Fraser). pp. 143-196. Martinus Nijhoff/Dr W. Junk, Dordrecht.

Fraser, R.S.S. (1986). Genes for resistance to plant viruses. CRC Crit. Rev. Plant Sci. 3, 257-294.

Fraser, R.S.S., and Clay, C.M. (1983). Pathogenesis-related proteins and acquired sysytemic resistance: causal relationship or separate effects? Neth. J. Plant Pathol. 89, 283-292.

Fraser, R.S.S., and Gerwitz, A. (1980). Tobacco mosaic virus infection does not alter the polyadenylated messenger RNA content of tobacco leaves. J. Gen. Virol. 46, 139-148.

Fraser, R.S.S., and Gerwitz, A. (1986). The genetics of resistance and virulence in plant virus diseases. In Genetics and Plant Pathogenesis. (Edited by P.R. Day and J.G. Jellis). Blackwells, Oxford, in press.

Fraser, R.S.S., Gerwitz, A., Loughlin, S.A.R., and Leary, J.A. (1983). Resistance to tobacco mosaic virus in tomato plants. Rep. Natl. Veg. Res. Stn. for 1982, 18-19.

Fraser, R.S.S., Gerwitz, A., and Morris, G.E.L. (1986a). Multiple regression analysis of the relationships between tobacco mosaic virus multiplication, the severity of mosaic symptoms, and the growth of tobacco and tomato. Physiol. Molec. Plant Pathol., in press.

Fraser, R.S.S., Gerwitz, A., Whenham, R.J., and Payne, J.A. (1986b). Virus infection, symptom development and control of plant growth. Rep. Natl. Veg. Res. Stn. for 1985, 23-25.

Fraser, R.S.S., Klöpfer, U., and Greenberg, D.A. (1973). The synthesis of TMV-RNA and ribosomal RNA in cultured tomato root tips. Virology 52, 275-280.

Fraser, R.S.S., and Loughlin, S.A.R. (1980). Resistance to tobacco mosaic virus in tomato: effects of the Tm-1 gene on virus multiplication. J. Gen. Virol. 48, 87-96.

Fraser, R.S.S., and Loughlin, S.A.R. (1982). Effects of temperature on the Tm-1 gene for resistance to tobacco mosaic virus in tomato. Physiol. Plant Pathol. 20, 109-117.

Fraser, R.S.S., Loughlin, S.A.R., and Whenham, R.J. (1979). Acquired systemic susceptibility to infection by tobacco mosaic virus in Nicotiana glutinosa L. J. Gen. Virol. 43, 131-141.

Fraser, R.S.S., and Whenham, R.J. (1978a). Chemotherapy of plant virus disease with methyl benzimidazole-2yl-carbamate: effects on plant growth and multiplication of tobacco mosaic virus. Physiol. Plant Pathol. 13, 51-64.

Fraser, R.S.S., and Whenham, R.J. (1978b). Inhibition of the multiplication of tobacco mosaic virus by methyl benzimidazol-2yl carbamate. J. Gen. Virol. 39, 191-194.

Fraser, R.S.S., and Whenham, R.J. (1982). Plant growth regulators and virus infection: a critical review. Plant Growth Regulation 1, 37-59.

Fritig, B., Gosse, J., Legrand, M., and Hirth, L. (1973). Changes in phenylalanine ammonia-lyase during the hypersensitive reaction of tobacco to TMV. Virology 55, 371-379.

Fritig, B., Legrand, M., and Hirth, L. (1972). Changes in the metabolism of phenolic compounds during the hypersensitive reaction of tobacco to TMV. Virology 47, 845-848.

Frosheiser, F.I. (1974). Alfalfa mosaic virus transmission to seed through alfalfa gametes and longevity in alfalfa seed. Phytopathology 64, 102-105.

Fukunaga, Y., Nagata, T., and Takebe, I. (1981). Liposome-mediated infection of plant protoplasts with tobacco mosaic virus. Virology 113, 752-760.

Fulton, R.W. (1951). Superinfection by strains of tobacco mosaic virus. Phytopathology 41, 579-592.

Fulton, R.W. (1978). Superinfection by strains of tobacco streak virus. Virology 85, 1-8.

Fulton, R.W. (1980). Biological significance of multicomponent viruses. Annu. Rev. Phytopathol. 18, 131-146.

Fulton, R.W. (1980). The protective effects of systemic virus infection. In Active Defense Mechanisms in Plants. (Edited by R.K.S. Wood). pp. 231-246. Plenum, New York.

Furusawa, I., and Okuno, T. (1978). Infection with BMV of mesophyll protoplasts from five plant species. J. Gen. Virol 40, 489-491.

Gaard, G., and De Zoeten, G.A. (1979). Plant virus uncoating as a result of virus-cell wall interactions. Virology 96, 21-31.

Gaborjanyi, R., Balazs, E., and Kiraly, Z. (1971). Ethylene production, tissue senescence and local virus infection. Acta Phytopathol. Acad. Sci. Hung. 6, 51-55.

Gangulee, R., Singh, B.R., and Singh, H.C. (1978). Studies on the metabolism of cowpea leaves infected with southern bean mosaic virus. 1. Effect on carbohydrate metabolism. Sci. Cult. 44, 226-228.

Gat-Edelbaum, O., Altman, A., and Sela, I. (1983). Polyinosinic:polycytidylic acid in association with cyclic nucleotides activates the antiviral factor AVF in plant tissues. J. Gen. Virol. 64, 211-214.

Gense, M.T. (1980a). DNA synthesis in wounded or TMV infected leaves from sensitive and hypersensitive tobacco plants. Cell Differ. 9, 117-124

Gense, M.T. (1980b). Heterogeneity of nuclear DNA labelling from intact, wounded and virus infected tobacco leaves. Cell Differ. 9, 125-133.

Gera, A., Loebenstein, G., and Shabtai, S. (1983). Enhanced tobacco mosaic virus production and suppressed synthesis of a virus inhibitor in protoplasts exposed to antibiotics. Virology 127, 475-478.

Gessner, S.L., and Irvin, J.D. (1980). Inhibition of elongation factor 2-dependent translocation by the pokeweed antiviral protein and ricin. J. Biol. Chem. 225, 3251-3253.

Ghabrial, S.A., and Lister, R.M. (1973). Coat protein and symptom specification in tobacco rattle virus. Virology 52, 1-12.

Ghorpade, L.N., and Joshi, G.V. (1980). Development of photosynthesis in the sugarcane plant (var. Co. 740) infected by mosaic virus. Indian J. Exp. Biol. 18, 1202-1203.

Ghosh, S.K. (1982). Growth promotion in plants by rice necrosis mosaic virus. Planta 155, 193-198.

Gianinazzi, S. (1982). Antiviral agents and inducers of virus resistance: analogies with interferon. In Active Defense Mechanisms in Plants. (Edited by R.K.S. Wood). pp. 275-298, Plenum Press, New York.

Gianinazzi, S., and Ahl, P. (1983). The genetic and molecular basis of b-proteins in the genus Nicotiana. Neth. J. Plant Pathol. 89, 275-281.

Gianinazzi, S., Ahl, P., Cornu, A., and Scalla, R. (1980). First report of host b-protein appearance in response to a fungal infection in tobacco. Physiol. Plant Pathol. 16, 337-342.

Gianinazzi, S., and Kassanis, B. (1974). Virus resistance induced in plants by polyacrylic acid. J. Gen. Virol. 23, 1-9.

Gianinazzi, S., Martin, C., and Valee, J.-C. (1970). Hypersensibilité aux virus, temperature et proteines solubles chez le Nicotiana Xanthi n.c. Apparition de nouvelles macromolecules lors de la repression de la synthèse virale. C. R. Hebd. Seances Acad. Sci. 270D, 2383-2386.

Gianinazzi, S., Pratt, H.M., Shewry, P.R., and Miflin, B.J. (1977). Partial purification and preliminary characterization of soluble leaf proteins specific to virus infected tobacco plants. J. Gen. Virol. 34, 345-351.

Gibbs, A. (1969). Plant virus classification. Adv. Virus Res. 14, 263-328.

Gibbs, A., and Harrison, B.D. (1976). Plant Virology: the Principles. Arnold, London. 292 pp.

Gibson, R.W., and Pickett, J.A. (1983). Wild potato repels aphids by release of aphid alarm pheromone. Nature 302, 608-609.

Gibson, R.W., and Plumb, R. (1977). Breeding for resistance to aphid infestation. In Aphids as Virus Vectors. (Edited by K.F. Harris and K. Maramorosch). pp. 473-500. Academic Press, New York.

Gill, C.C., and Chong, J. (1981). Vascular cell alterations and predisposed xylem infection in oats by inoculation with paired barley yellow dwarf isolates. Virology 114, 405-413.

Gill, D.S., Kumarasamy, R., and Symons, R.H. (1981). Cucumber mosaic virus-induced RNA replicase: solubilization and partial purification of the particulate enzyme. Virology 113, 1-8.

Gilpatrick, J.D., and Weintraub, M. (1952). An unusual type of protection with the carnation mosaic virus. Science 115, 701-702.

Goelet, P., Lomonossoff, G.P., Butler, P.J.G., Akam, M.E., Gait, M.J., and Karn, J. (1982). Nucleotide sequence of tobacco mosaic virus RNA. Proc. Natl. Acad. Sci. USA. 79, 5818-5822.

Goffeau, A, and Bove, J.M. (1965). Virus infection and photosynthesis. I. Increased photophosphorylation by chloroplasts from Chinese cabbage infected by turnip yellow mosaic virus. Virology 27, 243-252.

Goldbach, R., and Krijt, J. (1982). Cowpea mosaic virus-encoded protease does not recognize primary translation products of mRNAs from other comoviruses. J. Virol. 43, 1151-1154.

Goldbach, R.W., Rezelman, G., and Van Kammen, A. (1980). Independent replication and expression of the B-component RNA of cowpea mosaic virus. Nature 286, 297-299.

Goldbach, R.W., Schilthuis, J.G., and Rezelman, G. (1981). Comparison of in vivo and in vitro translation of cowpea mosaic virus RNAs. Biochem. Biophys. Res. Commun. 99, 89-94.

Gonda, T.J., and Symons, R.H. (1979). Cucumber mosaic virus replication in cowpea protoplasts: time course of virus, coat protein and RNA synthesis. J. Gen. Virol. 45, 723-736.

Goodman, T.C., Nagel, L., Rappoid, W., Klotz, G., and Reisner, D. (1984). Viroid replication: equilibrium association constant and comparative activity measurements for the viroid-polymerase interaction. Nucleic Acids Res. 12, 6231-6246.

Gordon, K.H.J., Gill, D.S., and Symons, R.A. (1982). Highly purified cucumber mosaic virus-induced RNA-dependent RNA polymerase does not contain any of the full length translation products of the genomic RNAs. Virology 123, 284-295.

Grasso, S., Jones, P., and White, R.F. (1980). Inhibition of tobacco mosaic virus multiplication in tobacco protoplasts by the pokeweed inhibitor. Phytopathol. Z. 98, 53-58.

Grasso, S., and Shepherd, R.J. (1978). Isolation and partial characterization of virus inhibitors from species taxonomically related to Phytolacca. Phytopathology 68, 199-205.

Gray, R.E., and Cashmore, A.R. (1976). RNA synthesis in plant leaf tissue: the characterization of messenger RNA species lacking and containing polyadenylic acid. J. Mol. Biol. 108, 595-608.

Grill, L.K. (1983). Utilizing RNA viruses for plant improvement. Plant Mol. Biol. Rep. 1, 17-20.

Grill, L.K., Garger, S.J., Turpen, T.H., Lowell, S.A., Marsden, M.P.E., and Murry, L.E. (1983). Involvement of viruses and virus-like agents with the male sterility trait of plants. In Plant Molecular Biology. (Edited by R.B. Goldberg). pp. 101-116. Alan R. Liss, Inc., New York.

Gross, H.J. (1985). Viroids: their structure and possible origin. In Subviral Pathogens of Plants and Animals: Viroids and Prions. (Edited by K. Maramorosch and J.J. McKelvey). pp. 165-182. Academic Press, Orlando.

Gross, H.J., Krupp, G., Domdey, H., Raba, M., Alberty, H., Lossow, C.H., Ramm, K., and Sänger, H.L. (1982). Nucleotide sequence and secondary structure of citrus exocortis and chrysanthemum stunt viroid. Eur. J. Biochem. 121, 249-257.

Gross, H.J., Liebl, U., Alberty, H., Krupp, G., Domdey, H., Ramm, K., and Sänger, H.L. (1981). A severe and mild potato spindle tuber viroid isolate differ in three nucleotide exchanges only. Biosci. Rep. 1, 235-241.

Guilfoyle, T.J. (1980). Transcription of cauliflower mosaic virus genome in isolated nuclei from turnip leaves. Virology 107, 71-80.

Guilfoyle, T.J., and Olszewski, N. (1983). The structure and transcription of the cauliflower mosaic virus minichromosome. In Plant Infectious Agents: Viruses, Viroids, Virusoids and Satellites. (Edited by H.D.Robertson, S.H. Howell, M. Zaitlin and R.L.Malmberg). pp. 34-38. Cold Spring Harbor Laboratory, New York.

Guilley, H., Richards, K.E., and Jonard, G. (1983). Observations concerning the discontinuous DNAs of cauliflower mosaic virus. EMBO J. 2, 277-282.

Haber, S., and Hamilton, R.I. (1980). Distribution of determinants for symptom production, nucleoprotein component distribution and antigenicity of coat protein between the two RNA components of cherry leaf roll virus. J. Gen. Virol. 50, 377-389.

Habili, N., and Francki, R.I.B. (1974). Comparative studies on tomato aspermy and cucumber mosaic virus. III. Further studies on relationships and construction of a virus from parts of the two viral genomes. Virology 61, 443-449.

Habili, N., and Kaper, J.M. (1981). Cucumber mosaic virus-associated

RNA 5. VII. Double-stranded form accumulation and disease attenuation in tobacco. Virology 112, 250-261.

Haehnel, W. (1984). Photosynthetic electron transport in higher plants. Annu. Rev. Plant Physiol. 35, 659-693.

Hall, A.E., and Loomis, R.S. (1972a). Photosynthesis and respiration by healthy and beet yellow virus-infected sugar beets (Beta vulgaris L.). Crop Sci. 12, 566-572.

Hall, A.E., and Loomis, R.S. (1972b). An explanation for the difference in photosynthetic capabilities of healthy and beet yellows infected sugar beets (Beta vulgaris L.). Plant Physiol. 50, 576-580.

Hall, H.K., and McWha, J.A. (1981). Effects of abscisic acid on growth of wheat (Triticum aestivum L.). Ann. Bot. 47, 427-433.

Hall, T.J. (1980). Resistance at the Tm-2 locus in tomato to tomato mosaic virus. Euphytica 29, 189-197.

Hamilton, R.I., and Dodds, J.A. (1970). Infection of barley by tobacco mosaic virus in single and mixed infection. Virology 42, 266-268.

Hamilton, R.I., and Nichols, C. (1977). The influence of bromegrass mosaic virus on the replication of tobacco mosaic virus in Hordeum vulgare. Phytopathology 67, 484-489.

Hampton, R.O. (1975). The nature of bean yield reduction by bean yellow and bean common mosaic viruses. Phytopathology 65, 1342-1346.

Hanada, K., and Harrison, B.D. (1977). Effects of virus genotype and temperature on seed transmission of nepoviruses. Ann. Appl. Biol. 85, 79-92.

Hari, V. (1980). Poly(A) polymerase activity in healthy and tobacco etch virus infected leaves. Phytopathol. Z. 99, 155-162.

Hariharasubramanian, V., Hadidi, A., Singer, B., and Fraenkel-Conrat, H. (1973). Possible identification of a protein in brome mosaic virus infected barley as a component of viral RNA polymerase. Virology 54, 190-198.

Harpaz, I., and Applebaum, S.W. (1961). Accumulation of asparagine in maize plants infected with maize rough dwarf virus and its significance in plant virology. Nature 192, 780-781.

Harrison, B.D., Murant, A.F., Mayo, M.A., and Roberts, I.M. (1974). Distribution of determinants for symptom production, host range and nematode transmissibility between the two RNA components of raspberry ringspot virus. J. Gen. Virol. 22, 233-247.

Hartung, W., Heilmann, B., and Gimmler, H. (1981). Do chloroplasts play a role in abscisic acid synthesis? Plant Sci. Lett. 22, 235-242.

Hatch, M.D. (1976). Photosynthesis - the path of carbon. In Plant Biochemistry. (Edited by J. Bonner and J.E.Varner). pp. 797-845. Academic Press, New York.

Hatta, T., and Matthews, R.E.F. (1974). The sequence of early cytological changes in Chinese cabbage leaf cells following systemic infection with turnip yellow mosaic virus. Virology 59, 383-396.

Hayashi, T. (1977). Fate of tobacco mosaic virus after entering the host cell. III. Partial uncoating. Microbiol. Immunol. 21,

317-324.
Hayashi, T., and Matsui, C. (1965). Fine structure of lesion periphery produced by tobacco mosaic virus. Phytopathology 55, 387-392.
Hayashi, Y. (1962). Amino acid activation in tobacco leaves infected with tobacco mosaic virus. Virology 18, 140-141.
Hebert, T.T. (1982). The rational for the Horsfall-Barratt disease assessment scale. Phytopathology 72, 1269.
Hecht, E.I., and Bateman, D.F. (1964). Nonspecific acquired resistance to pathogens resulting from localized infections by Thielaviopsis basicola or viruses in tobacco leaves. Phytopathology 54, 523-530.
Helms, K., and Wardlaw, I.F. (1978). Translocation of tobacco mosaic virus and photosynthetic assimilate in Nicotiana glutinosa. Physiol. Plant Pathol. 13, 23-36.
Helms, K., Waterhouse, P.M., and Mueller, W.J. (1985). Subterranean clover red leaf virus disease: effects of temperature on plant symptoms, growth and virus content. Phytopathology 75, 337-341.
Hennig, B., and Wittman, H.G. (1972). Tobacco mosaic virus: mutants and strains. In Principles and Techniques in Plant Virology. (Edited by C.I. Kado and H.O. Agrawal). pp. 546-594. Van Nostrand-Reinhold, New York.
Henry, E.W. (1983). Ultrastructural localization of tobacco mosaic virus in leaf tissue of Nicotiana tabacum L. cv. "Little Turkish". Cytobiol. 36, 7-16.
Hepburn, A.G., Wade, M., and Fraser, R.S.S. (1985). Present and future prospects for exploitation of resistance in crop protection by novel means. In Mechanisms of Resistance to Plant Diseases. (Edited by R.S.S. Fraser). pp. 425-452. Martinus Nijhoff/Dr W. Junk, Dordrecht.
Hiebert, E., Bancroft, J.B., and Bracker, C.E. (1968). The assembly in vitro of some small spherical viruses, hybrid viruses and other nucleoproteins. Virology 34, 492-508.
Higgins, T.J.V., Goodwin, P.B., and Whitfield, P.R. (1976). Occurrence of short particles in beans infected with the cowpea strain of TMV. II. Evidence that short particles contain the cistron for coat protein. Virology 71, 486-497.
Hillman, B.I., Morris, T.J., and Schlegel, D.E. (1985). Effects of low-molecular-weight RNA and temperature on tomato bushy stunt virus symptom expression. Phytopathology 75, 361-365.
Hirai, A., and Wildman, S.G. (1969). Effect of TMV multiplication on RNA and protein synthesis in tobacco chloroplasts. Virology 38, 73-82.
Hohn, T., Pietrzak, M., Dixon, L., Koenig, I., Penswick, J., Hohn, B., and Pfeiffer, P. (1983). Involvement of reverse transcription in cauliflower mosaic virus replication. In Plant Infectious Agents. (Edited by H.D. Robertson, S.H. Howell, M. Zaitlin and R. L. Malmberg). pp. 28-33. Cold Spring Harbor Laboratory, New York.
Hollings, M., and Brunt, A.A. (1981). Potyviruses. In Handbook of Plant Virus Infections: Comparative Diagnosis. (Edited by E. Kurstak). pp. 732-807. Elsevier/North Holland, Amsterdam.
Holmes, F.O. (1938). Inheritance of resistance to tobacco mosaic

disease in tobacco. Phytopathology 28, 553-561.
Holmes, F.O. (1955). Additive resistances to specific viral diseases in plants. Ann. Appl. Biol. 42, 129-139.
Hooft van Huijsduijnen, R.A.M., Cornelissen, B.J.C., Van Loon, L.C., Van Boom, J.H., Tromp, M., and Bol, J.F. (1985). Virus-induced synthesis of messenger RNAs for precursors of pathogenesis-related proteins in tobacco. EMBO J. 4, 2167-2171.
Hooley, R., and McCarthy, D. (1980). Extracts from virus infected hypersensitive tobacco leaves are deterimental to protoplast survival. Physiol. Plant Pathol. 16, 25-38.
Hsu, C.H., Sehgal, O.P., and Pickett, E.E. (1976). Stabilizing effect of divalent metal ions on virions of southern bean mosaic virus. Virology 69, 587-595.
Huber, R., Hontilez, J., and Van Kammen, A. (1981). Infection of cowpea protoplasts with both the common strain and the cowpea strain of TMV. J. Gen. Virol. 55, 241-245.
Huber, R., Rezelman, G., Hibi, T., and Van Kammen, A. (1977). Cowpea mosaic virus infection of protoplasts from Samsun tobacco leaves. J. Gen. Virol. 34, 315-323.
Huisman, M.J., Broxterman, H.J.G., Schellekens, H., and Van Vloten-Doting, L. (1985). Human interferon does not protect cowpea plant cell protoplasts against infection with alfalfa mosaic virus. Virology 143, 622-625.
Hull, R. (1978). The stabilization of the particles of turnip rosette virus. III. Divalent cations. Virology 89, 418-422.
Hull, R., and Davies, J.W. (1983). Genetic engineering with plant viruses and their potential as vectors. Adv. Virus Res. 28, 1-35.
Hull, R., and Covey, S.N. (1983). Does cauliflower mosaic virus replicate by reverse transcription? Trends Biochem. Sci. 8, 119-121.
Hunter, C.S., and Peat, W.E. (1973). The effect of tomato aspermy virus on photosynthesis in the young tomato plant. Physiol. Plant Pathol. 3, 517-524.
Ikegami, M., and Fraenkel Conrat, H. (1978a). The RNA-dependent RNA polymerase of cowpea. FEBS Lett. 96, 197-200.
Ikegami, M., and Fraenkel Conrat, H. (1978b). RNA-dependent RNA polymerase of tobacco plants. Proc. Natl. Acad. Sci. USA. 75, 2122-2124.
Ikegami, M., and Fraenkel Conrat, H. (1979). Characterization of double-stranded ribonucleic acid in tobacco leaves. Proc. Natl. Acad. Sci. USA. 76, 3637-3640.
Irvin, J.D., and Aron, G.M. (1982). Chemical modifications of pokeweed antiviral protein: effects upon ribosome inactivation, antiviral activity and cytotoxicity. FEBS Lett. 148, 127-130.
Irvin, J.D., Kelly, T., and Robertus, J.D. (1980). Purification and properties of a second antiviral protein from *Phytolacca americana* which inactivates eukaryotic ribosomes. Arch. Biochem. Biophys 20, 418-425.
Izant, J.G., and Weintraub, H. (1985). Constitutive and conditional suppression of exogenous and endogenous genes by anti-sense RNA. Science 229, 345-352.
Jackson, A.O., Mitchell, D.M., and Siegel, A. (1971). Replication of

tobacco mosaic virus. I. Isolation and characterization of double-stranded forms of ribonucleic acid. Virology 45, 182-191.

Jaspars, E.M.J., Gill, D.S., and Symons, R.H. (1985). Viral RNA synthesis by a particulate fraction from cucumber seedlings infected with cucumber mosaic virus. Virology 144, 410-425.

Jedlinski, H., Rochow, W.F., and Brown, C.M. (1977). Tolerance to barley yellow dwarf virus in oats. Phytopathology 67, 1408-1411.

Jensen, S.G. (1968). Photosynthesis, respiration and other physiological relationships in barley infected with barley yellow dwarf virus. Phytopathology 58, 204-208.

Jensen, S.G. (1969a). Composition and metabolism of barley leaves infected with barley yellow dwarf virus. Phytopathology 59, 1694-1698.

Jensen, S.G. (1969b). Occurrence of virus particles in the phloem tissue of BYDV-infected barley. Virology 38, 83-91.

Jensen, S.G. (1972). Metabolism and carbohydrate composition in barley yellow dwarf virus-infected wheat. Phytopathology 62, 587-592.

Jockusch, H. (1966a). Temperatursensitive Mutanten des Tabakmosaikvirus. II. In vitro Verhalten. Z. Vererbungsl. 98, 344-362.

Jockusch, H. (1966b). Temperatursensitive Mutanten des Tabakmosaikvirus. I. In vivo Verhalten. Z. Vererbungsl. 98, 320-343.

Jockusch, H. (1968). Two mutants of tobacco mosaic virus temperature-sensitive in two different functions. Virology 35, 94-101.

Jockusch, H. (1974). Zellfreie Bildung eines Replikationsenzyms. Naturwissenschaften 61, 267-269.

Jockusch, H., and Jockusch, B. (1968). Early cell death caused by TMV-mutants with defective coat proteins. Mol. Gen. Genet. 102, 204-209.

John, V.T., and Weintraub, M. (1967). Phenolase activity in Nicotiana glutinosa infected with tobacco mosaic virus. Phytopathology 57, 154-158.

Johnson, C.S., Main, C.E., and Gooding, G.V. (1983). Crop loss assessment for flue-cured tobacco cultivars infected with tobacco mosaic virus. Plant Dis. 67, 881-885.

Jones, A.T., and Catherall, P.L. (1970). The relationship between growth rate and the expression of tolerance to barley yellow dwarf virus in barley. Ann. Appl. Biol. 65, 137-145.

Jones, A.T., and Duncan, G.H. (1980). The distribution of some genetic determinants in the two nucleoprotein particles of cherry leaf roll virus. J. Gen. Virol. 50, 269-277.

Jones, I.M., and Reichmann, M.E. (1973). The proteins synthesized in tobacco leaves infected with tobacco necrosis virus and satellite tobacco necrosis virus. Virology 52, 49-56.

Jons, V.L., Timian, R.G., and Lamey, H.A. (1981). Effect of wheat streak mosaic virus on twelve hard red spring cultivars. N. D. Farm Res. 39, 17-18.

Kacser, H., and Burns, J.A. (1981). The molecular basis of dominance. Genetics 97, 639-666.

Kado, C.I., and Knight, C.A. (1966). Location of a local lesion gene in tobacco mosaic virus RNA. Proc. Natl. Acad. Sci. USA. 55, 1276-1283.

Kamen, R.I. (1975). Structure and functions of the Qβ RNA replicase. In RNA Phages. (Edited by N.D. Zinder). pp. 203-234. Cold Spring Harbor Laboratory, New York.

Kamer, G., and Argos, P. (1984). Primary structural comparison of RNA-dependent polymerases from plant, animal and bacterial viruses. Nucleic Acids Res. 12, 7269-7282.

Kano, H. (1985). Effects of light and inhibitors of photosynthesis and respiration on the multiplication of tobacco mosaic virus in tobacco protoplasts. Plant Cell Physiol. 26, 1241-1249.

Kaper, J.M., Tousignant, M.E., and Thompson, S.M. (1981). Cucumber mosaic virus-associated RNA 5. VII. Identification and partial characterization of a CARNA 5 incapable of inducing tomato necrosis. Virology 114, 526-533.

Kasamo, K., and Shimomura, T. (1978). Response of membrane-bound Mg^{2+}-activated ATPase of tobacco leaves to tobacco mosaic virus. Plant Physiol. 62, 731-734.

Kassanis, B. (1981). Some speculations on the nature of the natural defence mechanism of plants against virus infection. Phytopathol. Z. 102, 277-291.

Kassanis, B., Gianinazzi, S., and White, R.F. (1974). A possible explanation of the resistance of virus-infected tobacco plants to second infection. J. Gen. Virol. 23, 11-16.

Kassanis, B., and Kenten, R.H. (1978). Inactivation and uncoating of TMV on the surface and in the intercellular spaces of leaves. Phytopathol. Z. 91, 329-339.

Kassanis, B., and White, R.F. (1978). Effect of polyacrylic acid and b proteins on TMV multiplication in tobacco protoplasts. Phytopathol. Z. 91, 269-272.

Khadhair, A.H., Sinha, R.C., and Peterson, J.F. (1984). Effect of white clover mosaic virus infection on various processes relevant to symbiotic N_2 fixation in red clover. Can. J. Bot. 62, 38-43.

Kiho, Y., Machida, H., and Oshima, N. (1972). Mechanism determining the host specificity of tobacco mosaic virus. I. Formation of polysomes containing infecting viral genome in various plants. Japan J. Microbiol. 16, 451-459.

Kiho, Y., and Nishiguchi, M. (1984). Unique nature of an attenuated strain of tobacco mosaic virus: autoregulation. Microbiol. Immunol. 28, 589-599.

Kiho, Y., Shimomura, T., Abe, T., and Nozu, Y. (1979). Infectivity suppressing and virus-binding activities of a membrane material isolated from tobacco leaves. Microbiol. Immunol. 23, 735-748.

Kim, K.S. (1970). Subcellular responses to localized infection of Chenopodium quinoa by pokeweed mosaic virus. Virology 41, 179-183.

Kimmins, W.C. (1969). Isolation of a virus inhibitor from plants with localized infections. Can. J. Bot. 47, 1879-1886.

Kingsland, G.C. (1980). Effect of maize dwarf mosaic virus infection on yield and stalk strength of corn in the field in South Carolina. Plant Dis. 64, 271-273.

Kiss, T., Posfai, J., and Solymosy, F. (1983). Sequence homology between potato spindle tuber viroid and U3B snRNA. FEBS Lett. 163, 217-220.

Kluge, S. (1976). Proteingehalt in hell- und dunkelgrünen Blattbezirken mosaikkranker Tabakpflanzen. Biochem. Physiol. Pflanzen 170, 91-95.

Konate, G., Kopp, M., and Fritig, B. (1983). Studies on TMV multiplication in systemically and hypersensitively reacting tobacco varieties by means of radiochemical and immunoenzymatic methods. Agronomie 3, 95.

Kooistra, E. (1968). Significance of the non-appearance of visible disease symptoms in cucumber (Cucumis sativus L.) after infection with Cucumis virus 2. Euphytica 17, 136-140.

Kozak, M. (1979). Inability of circular mRNA to attach to eukaryotic ribosomes. Nature 280, 82-85.

Kraev, V.G., and Diden, L.F. (1982). Effect of cycloheximide and chloramphenicol on reproduction of potato virus X in isolated leaves of Datura stramonium. Mikrobiol. Zh. 44, 60-63.

Kubo, S. (1966). Chromatographic studies of RNA synthesis in tobacco leaf tissues infected with tobacco mosaic virus. Virology 28, 229-235.

Kubo, S., and Takanami, I. (1979). Infection of tobacco mesophyll protoplasts with tobacco necrotic dwarf virus, a phloem-limited virus. J. Gen. Virol. 42, 387-398.

Kubo, S., and Tomaru, K. (1968). Synthesis of virus and host nucleoproteins in tobacco mosaic virus-infected tobacco leaf tissue. In Biochemical Regulation in Diseased Plants or Injury. pp. 35-48. Phytochemical Society of Japan, Tokyo.

Kumarasamy, R., and Symons, R.H. (1979). Extensive purification of the cucumber mosaic virus-induced RNA replicase. Virology 96, 622-632.

Kuriger, W.E., and Agrios, G.N. (1977). Cytokinin levels and kinetin-virus interactions in tobacco ringspot virus infected cowpea plants. Phytopathology 67, 604-609.

Kurstak, E. (1981). (Ed.) Handbook of Plant Virus Infections and Comparative Diagnosis. Elsevier/North Holland, Amsterdam. 943 pp.

Lafleche, D., and Bove, J.M. (1971). Virus de la mosaique jaune du navet: site cellulaire de la replication du RNA viral. Physiol. Veg. 9, 487-503.

Lane, L.C. (1981). Bromoviruses. In Handbook of Plant Virus Infection: Comparative Diagnosis. (Edited by E. Kurstak). pp. 333-376. Elsevier/North Holland, Amsterdam.

Lastra, R., and Gil, F. (1981). Ultrastructural host cell changes associated with tomato yellow mosaic virus. Phytopathology 71, 524-528.

Latorre, B.A., Flores, V., and Marholz, G. (1984). Effect of potato virus Y on growth, yield and chemical composition of flue-cured tobacco in Chile. Plant Dis. 68, 884-886.

Leal, N., and Lastra, R. (1984). Altered metabolism of tomato plants infected with tomato yellow mosaic virus. Physiol. Plant Pathol. 24, 1-8.

Legrand, M, Fritig, B., and Hirth, L. (1978). o-Diphenol O-methyltransferases of healthy and tobacco mosaic virus-infected hypersensitive tobacco. Planta 144, 101-108.

Levy, A., Loebenstein, G., Smookler, M., and Drovi, T. (1974). Partial suppression by UV irradiation of the mechanism of resistance to cucumber mosaic virus in a resistant cucumber cultivar. Virology 60, 37-44.

Levy, D., and Marco, S. (1976). Involvement of ethylene in epinasty of CMV-infected cucumber cotyledons which exhibit increased resistance to gaseous diffusion. Physiol. Plant Pathol. 9, 121-126.

Lieberman, M. (1979). Biosynthesis and action of ethylene. Annu. Rev. Plant Physiol. 30, 533-591.

Lindsey, D.W., and Gudauskas, R.T. (1975). Effects of maize dwarf mosaic virus on water relations of corn. Phytopathology 65, 434-440.

Lisa, V., Luisoni, E., and Milne, R.G. (1981). A possible virus cryptic in carnation. Ann. Appl. Biol. 98, 431-437.

Lockhart, B.E.L., and Semancik. J.S. (1969). Differential effect of actinomycin D on plant virus multiplication. Virology 39, 362-365.

Lockhart, B.E.L., and Semancik, J.S. (1970). Growth inhibition, peroxidase and 3-indole-acetic acid oxidase activity, and ethylene production of cowpea mosaic virus-infected cowpea seedlings. Phytopathology 60, 553-556.

Loebenstein, G., Chazan, R., and Eisenberg, M. (1970). Partial suppression of the localizing mechanism to tobacco mosaic virus by uv irradiation. Virology 41, 373-376.

Loebenstein, G., Cohen, J., Shabtai, S., Coutts, R.H.A., and Wood, K.R. (1977). Distribution of cucumber mosaic virus in systemically infected tobacco leaves. Virology 81, 117-125.

Loebenstein, G., and Gera, A. (1981). Inhibitor of virus replication released from tobacco mosaic virus-infected protoplasts of a local lesion-responding tobacco cultivar. Virology 114, 132-139.

Loebenstein, G., Gera, A., Barnett, A., Shabtai, S., and Cohen, J. (1980). Effect of 2,4-dichlorophenoxyacetic acid on multiplication of TMV in protoplasts from local lesion and systemic responding hosts. Virology 100, 110-115.

Loebenstein, G., Rabina, S., and Van Praagh, T. (1968). Sensitivity of induced localized acquired resistance to actinomycin D. Virology 34, 264-268.

Loebenstein, G., and Ross, A.F. (1963). An extractable agent, induced in uninfected tissues by localized virus infections, that interferes with infection by tobacco mosaic virus. Virology 20, 507-517.

Loebenstein, G., Sela, I., and Van Praagh, T. (1969). Increase of tobacco mosaic virus local lesion size and virus multiplication in hypersensitive hosts in the presence of Actinomycin-D. Virology 37, 42-48.

Loesch, L.S., and Fulton, R.W. (1975). Prunus necrotic ringspot virus as a multicomponent system. Virology 68, 71-78.

Loesch-Fries, L.S., Halk, E.L., Nelson, S.E., and Krahn, K.J. (1985). Human leukocyte interferon does not inhibit alfalfa mosaic virus in protoplasts or tobacco tissue. Virology 143, 626-629.

Louie, R., and Darrah, L.L. (1980). Disease resistance and yield loss to sugarcane mosaic virus in East African-adapted maize. Crop Sci. 20, 638-640.

Loveys, B.R. (1977). The intracellular location of abscisic acid in stressed and non-stressed leaf tissue. Physiol. Plant. 40, 6-10.

Lucas, J., Camacho Henriquez, A., Lottspeich, F., Henschen, A.S., and Sänger, H.L. (1985). Amino acid sequence of the 'pathogenesis-related' protein p14 from viroid-infected tomato reveals a new type of structurally unfamiliar proteins. EMBO J. 4, 2745-2749.

McCarthy, D., Bleichmann, S., and Thorne, J. (1980). Some effects of pH, salt, urea, ethanediol and sodium dodecyl sulphate on tobacco necrosis virus. J. Gen. Virol. 46, 391-404.

McCarthy, D., Jarvis, B.C., and Thomas, B.J. (1970). Changes in the ribosomes extracted from mung beans infected with a strain of tobacco mosaic virus. J. Gen. Virol. 9, 9-17.

McClintock, B. (1978). Mechanisms that rapidly reorganize the genome. Stadler Genet. Symp. 10, 25-47.

McIntyre, J.L., Dodds, J.A., and Hare, J.D. (1981). Effects of localized infections of Nicotiana tabacum by tobacco mosaic virus on systemic resistance against diverse pathogens and an insect. Phytopathology 71, 297-301.

MacKenzie, D.R. (1983). Toward the management of crop losses. In Challenging Problems in Plant Health. (Edited by T. Kommendahl and P.H. Williams). pp. 82-92. American Phytopathological Society, St. Paul.

McKinney, H.H. (1929). Mosaic diseases in the Canary Islands, West Africa and Gibraltar. J. Agric. Res. 39, 557-578.

McRitchie, J.J., and Alexander, L.J. (1963). Host-specific Lycopersicon strains of tobacco mosaic virus. Phytopathology 53, 394-398.

Maekawa, K., Furusawa, I., and Okuno, T. (1981). Effects of actinomycin-D and ultraviolet irradiation on multiplication of brome mosaic virus in host and non-host cells. J. Gen. Virol. 53, 353-356.

Magyarosy, A.C., Buchanan, B.B., and Schürmann, P. (1973). Effect of a systemic virus infection on chloroplast function and structure. Virology 55, 426-438.

Makovcova, O., and Sindelar, L. (1977). Changes in glycolysis and pentose phosphate cycle activity with TMV infected tobacco. Biol. Plant. 19, 253-258.

Makovcova, O., and Sindelar, L. (1978). Changes in phosphoenolpyruvate carboxylase and ribulosebisphosphate carboxylase activities in tobacco plants infected with tobacco mosaic virus. Biol. Plant. 20, 135-137.

Manners, J.M., and Scott, K.J. (1985). Reduced translatable messenger RNA activities in leaves of barley infected with Erysiphe graminis f. sp. hordei. Physiol. Plant Pathol. 26, 297-308.

Marco, S., and Levy, D. (1979). Involvement of ethylene in the development of cucumber mosaic virus-induced chlorotic lesions in cucumber cotyledons. Physiol. Plant Pathol. 14, 235-244.

Marco, S., Levy, D., and Aharoni, N. (1976). Involvement of ethylene in the suppression of hypocotyl elongation in CMV-infected

cucumbers. Physiol. Plant Pathol. 8, 1-7.
Martin, C., and Martin-Tanguy, J. (1981). Polyamines conjugées et limitation de l'expansion virale chez lesvégétaux. C. R. Hebd. Seances Acad. Sci. 293, 249-251.
Martin-Tanguy, J., Martin, C., Gallet, M., and Vernoy, R. (1976). Sur despuissants inhibiteurs naturels de multiplication du virus de la mosaique du tabac. C. R. Hebd. Seances Acad. Sci. 282, 2231-2234.
Massala, R., Legrande, M., and Fritig, B. (1980). Effect of α-aminoacetate, a competitive inhibitor of phenylalanine ammonia lyase, on the hypersensitive resistance of tobacco to tobacco mosaic virus. Physiol. Plant Pathol. 16, 213-226.
Matkovics, B., Novak, R., Szabo, L., Varga, S.I., and Farkas, G.J. (1978). Enzymatic and metabolic changes of tomato plants after TMV infection. Biochem. Physiol. Pflanzen 172, 315-318.
Matsuoka, M., and Ohashi, Y. (1986). Induction of pathogenesis-related proteins in tobacco leaves. Plant Physiol. 80, 505-510.
Matthews, R.E.F. (1973). Induction of disease by viruses, with special reference to turnip yellow mosaic virus. Annu. Rev. Phytopathol. 11, 147-170.
Matthews, R.E.F. (1979). The classification and nomenclature of viruses. Intervirology 11, 133-135.
Matthews, R.E.F. (1981). Plant Virology. Second Edition. Academic Press, New York. 897 pp.
Matthews, R.E.F., and Sarkar, S. (1976). A light induced structural change in chloroplasts of Chinese cabbage cells infected with turnip yellow mosaic virus. J. Gen. Virol. 33, 435-446.
Matthews, R.E.F., and Witz, J. (1985). Uncoating of turnip yellow mosaic virus RNA in vivo. Virology 144, 318-327.
Maule, A.J., Bouton, M.I., and Wood, K.R. (1980). Resistance of cucumber protoplasts to cucumber mosaic virus: a comparative study. J. Gen. Virol. 51, 271-279.
May, J.T., Gilliland, J.M., and Symons, R.H. (1969). Plant virus-induced RNA polymerase: properties of the enzyme partly purified from cucumber cotyledons infected with cucumber mosaic virus. Virology 39, 54-65.
Mayo, M.A., and Barker, H. (1983). Effects of actinomycin-D on the infection of tobacco protoplasts by four viruses. J. Gen. Virol. 64, 1775-1780.
Menissier, J., Pfeiffer, P, Lebeurier, G., Guilley, H., Lacroute, F., and Hirth, L. (1983). Cauliflower mosaic virus DNA in the elaboration of gene vectors in plants. In Plant Infectious Agents. (Edited by H.D Robertson, S.H. Howell, M. Zaitlin. and R.L. Malmberg). pp. 44-48. Cold Spring Harbor Laboratory, New York.
Menke, G.H., and Walker, J.K. (1963). Metabolism of resistant and susceptible cucumber varieties infected with cucumber mosaic virus. Phytopathology 53, 1349-1355.
Meshi, T., Ishikawa, M., Ohno, T., Okada, Y, Sano, T., Ueda, I., and Shikata, E. (1984). Double-stranded cDNAs of hop stunt viroid are infectious. J. Biochem. Tokyo 95, 1521-1524.
Milborrow, B.V. (1974). Biosynthesis of abscisic acid by a cell-free system. Phytochemistry 13, 131-136.

Miller, W.A., Dreher, T.W., and Hall, T.C. (1985). Synthesis of brome mosaic virus subgenomic RNA in vitro by internal initiation on (-) sense genomic RNA. Nature 313, 68-70.

Miller, W.A., and Hall, T.C. (1983). Use of micrococcal nuclease in the purification of highly template dependent RNA-dependent RNA polymerase from brome mosaic virus-infected barley. Virology 125, 236-241.

Milne, R.G. (1966). Multiplication of tobacco mosaic virus in tobacco leaf palisade cells. Virology 28, 79-89.

Milner, J.J., and Jackson, A.O. (1983). Characterization of viral-complementary RNA associated with polyribosomes from tobacco infected with sonchus yellow net virus. J. Gen. Virol. 64, 2479-2483.

Modderman, P.W., Schot, C.P., Klis, F.M., and Wieringa-Brants, D.H. (1985). Acquired resistance in hypersensitive tobacco against tobacco mosaic virus induced by plant cell wall components. Phytopathol. Z. 113, 165-170.

Mohammed, N.A. (1973). Some effects of systemic infection by tomato spotted wilt virus on chloroplasts of Nicotiana tabacum leaves. Physiol. Plant Pathol. 3, 509-516.

Mohammed, N.A., and Randles, J.W. (1972). Effect of tomato spotted wilt virus on ribosomes, ribonucleic acids and fraction 1 protein in Nicotiana tabacum leaves. Physiol. Plant Pathol. 2, 235-245.

Mohanty, S.K., and Sridhar, R. (1982). Physiology of rice tungro virus disease: proline accumulation due to infection. Physiol. Plant. 56, 89-93.

Motoyoshi, F., Bancroft, J.B., and Watts, J.W. (1974). The infection of tobacco protoplasts with a variant of brome mosaic virus. J. Gen. Virol. 25, 31-36.

Motoyoshi, F., and Oshima, N. (1977). Expression of genetically controlled resistance to tobacco mosaic virus infection in isolated tomato leaf mesophyll protoplasts. J. Gen. Virol. 34, 499-506.

Motoyoshi, F., and Oshima, N. (1979). Standardization in inoculation procedure and effect of a resistance gene on infection of tomato protoplasts with tobacco mosaic virus RNA. J. Gen. Virol. 44, 801-806.

Mottinger, J.P., and Dellaporta, S.L. (1983). Stable and unstable mutations associated with virus infection in maize. In Plant Infectious Agents: Viruses, Viroids, Virusoids and Satellites. (Edited by H.D.Robertson, S.H. Howell, M. Zaitlin and R.L.Malmberg). pp. 126-129. Cold Spring Harbor Laboratory, New York.

Mouches, C., Bove, C., and Bove, J.M. (1974). Turnip yellow mosaic virus-RNA replicase: partial purification of the enzyme from the solubilized enzyme-template complex. Virology 58, 409-423.

Mouches, C., Candresse, T., and Bove, J.M. (1984). Turnip yellow mosaic virus RNA-replicase contains host and virus-encoded subunits. Virology 134, 78-90.

Mount, S.M., Petterson, I., Hinterberger, M., Karmas, A., and Steitz, J. (1983). The U1 small nuclear RNA-protein complex selectively binds a 5' splice site in vitro. Cell 33, 509-518.

Mozes, R., Antignus, Y., Sela, I., and Harpaz, I. (1978). The chemical nature of an antiviral factor (AVF) from virus-infected plants. J. Gen. Virol. 38, 241-249.

Mühlbach, H.P., and Sänger, H.L. (1979). Viroid replication is inhibited by α-amanitin. Nature 278, 185-187.

Mundry, K.-W. (1963). Plant virus-host cell relationships. Annu. Rev. Phytopathol. 1, 173-195.

Murakishi, H.H., and Carlson, P.S. (1976). Regeneration of virus-free plants from dark green islands of tobacco mosaic virus-infected tobacco leaves. Phytopathology 66, 931-932.

Murant, A.F., Taylor, C.E., and Chambers, J. (1968). Properties, relationships and transmission of a strain of raspberry ringspot virus infecting raspberry cultivars immune to the common Scottish strain. Ann. Appl. Biol. 61, 175-186.

Nachman, I., Loebenstein, G., van Praagh, T., and Zelcer, A. (1971). Increased multiplication of cucumber mosaic virus in a resistant cucumber cultivar caused by actinomycin D. Physiol. Plant Pathol. 1, 67-71.

Naidu, R.A., Krishnan, M., Nayudu, M.V., and Gnanam, A. (1984a). Studies on peanut green mosaic virus infected peanut (Arachis hypogaea L.) leaves. II. Chlorophyll-protein complexes and polypeptide composition of thylakoid membranes. Physiol. Plant Pathol. 25, 191-198.

Naidu, R.A., Krishnan, M., Nayudu, M.V., and Gnanam, A. (1986). Studies on peanut green mosaic virus infected peanut (Arachis hypogaea L.) leaves. III. Changes in the polypeptides of photosystem II particles. Physiol. Molec. Plant Pathol., in press.

Naidu, R.A., Krishnan, M., Ramanujam, P., Gnanam, A., and Nayudu, M.V. (1984b). Studies on peanut green mosaic virus infected peanut (Arachis hypogaea L.) leaves. I. Photosynthesis and photochemical reactions. Physiol. Plant Pathol. 25, 181-190.

Nakagaki, Y., Hirai, T., and Stahmann, M.A. (1970). Ethylene production by detached leaves infected with tobacco mosaic virus. Virology 40, 1-9.

Nassuth, A., Alblas, F., Van der Geest, A.J.M., and Bol, J.F. (1983a). Inhibition of alfalfa mosaic virus RNA and protein synthesis by actinomycin D and cycloheximide. Virology 126, 517-524.

Nassuth, A.F., and Bol, J.F. (1983). Altered balance of the synthesis of plus- and minus-strand RNAs induced by RNAs 1 and 2 of alfalfa mosaic virus in the absence of RNA-3. Virology 124, 75-85.

Nassuth, A., ten Bruggencate, G., and Bol, J. (1983b). Time course of alfalfa mosaic virus RNA and coat protein synthesis in cowpea protoplasts. Virology 125, 75-84.

Nault, L.R., and Montgomery, M.E. (1977). Aphid pheromones. In Aphids as Virus Vectors. (Edited by K.F.Harris and K. Maramorosch). pp. 528-546. Academic Press, New York and London.

Nhu, K.M., Mayee, C.D., and Sarkar, S. (1982). Reduced synthesis of PVY in PVX-infected tobacco plants. Naturwissenschaften 69, 183-184.

Niblett, C.L., Dickson, E., Fernow, K.H., Horst, R.K., and Zaitlin, M. (1978). Cross protection among four viroids. Virology 91,

198-203.

Nienhaus, F., and Janicka-Czarnecka, I. (1981). Investigations on an extractable induced antiviral principle (AVP) in systemically diseased Phaseolus vulgaris upon tobacco mosaic virus infection. Z. Pflanzenkr. Pflanzenschutz 88, 577-583.

Nishiguchi, M., Kikuchi, S., Kiho, Y., Ohno, T., Meshi, T., and Okada, Y. (1985). Molecular basis of plant viral virulence; the complete nucleotide sequence of an attenuated strain of tobacco mosaic virus. Nucleic Acids Res. 13, 5585-5590.

Nishiguchi, M., Motoyoshi, F., and Oshima, N. (1978). Behaviour of a temperature sensitive strain of tobacco mosaic virus in tomato leaves and protoplasts. J. Gen. Virol. 39, 53-62.

Nishiguchi, M., Motoyoshi, F., and Oshima, N. (1980). Further investigation of a temperature-sensitive strain of tobacco mosaic virus: its behaviour in tomato leaf epidermis. J. Gen. Virol. 46, 497-500.

Nishiguchi, M., Nozy, Y., and Oshima, N. (1984). Comparative study on coat proteins of an attenuated, a temperature sensitive and their parent strains of tobacco mosaic virus. Phytopathol. Z. 109, 104-114.

Nitzany, F.E., and Cohen, S. (1960). A case of interference between alfalfa mosaic virus and cucumber mosaic virus. Virology 11, 771-773.

Novacky, A., and Hampton, R.E. (1968). Peroxidase isoenzymes in virus-infected plants. Phytopathology 58, 301-305.

Novak, J, (1975). Wachstumsänderungen der Tabakpflanze Nicotiana tabacum cv. Samsun nach der Inokulation des Tabakmosaikvirus. Acta Scien. Nat. Acad. Sci. Bohemoslov. Brno. 9, 1-41.

Novikov, V.K., and Atabekov, J.G. (1970). A study of the mechanisms controlling the host range of plant viruses. I. Virus specific receptors of Chenopodium amaranticolor. Virology 41, 101-107.

Nuss, D.L., and Peterson, A.J. (1981). Resolution and genome assignments of mRNA transcripts synthesized in vitro by wound tumour virus. Virology 114, 399-404.

O'Farrell, P.H. (1975). High resolution two-dimensional electrophoresis of proteins. J. Biol. Chem. 256, 4007-4021.

Ogren, W.L. (1984). Photorespiration: pathways, regulation and modification. Annu. Rev. Plant Physiol. 35, 415-442.

Ohno, T., Takamatsu, N., Meshi, T., Okada, Y., Nishiguchi, M., and Kiho, Y. (1983). Single amino acid substitution in 30 K protein of TMV defective in transport function. Virology 131, 255-258.

Ohashi, Y., and Shimomura, T. (1982). Modification of cell membranes of leaves systemically infected with tobacco mosaic virus. Physiol. Plant Pathol. 20, 125-128.

Olszewski, N., Hagen, G., and Guilfoyle, T. (1982). A transcriptionally active minichromosome of cauliflower mosaic virus DNA isolated from infected turnip leaves. Cell 29, 395-402.

Orchansky, P., Rubinstein, M., and Sela, I. (1982). Human interferons protect plants from virus infection. Proc. Natl. Acad. Sci. USA. 79, 2278-2280.

Orellana, R.G., Fan, F.F., and Sloger, C. (1982). Tobacco ringspot virus and Rhizobium japonicum interactions in soybean: impairment of leghemoglobin accumulation and nitrogen fixation. Phytopathology 68, 577-582.

Orellana, R.G., Weber, D.F., and Cregan, P. (1980). N_2-fixing competence of Rhizobium japonicum strains in soybean infected with tobacco ringspot virus. Physiol. Plant Pathol. 17, 381-388.

Otsuki, Y., Shimomura, T., and Takebe, I. (1972). Tobacco mosaic virus multiplication and expression of the N gene in necrotic responding tobacco varieties. Virology 50, 45-50.

Owens, R.A., and Bruening, G. (1975). The pattern of amino acid incorporation into two cowpea mosaic virus proteins in the presence of ribosome-specific protein synthesis inhibitors. Virology 64, 520-530.

Owens, R.A., Bruening, G., and Shepherd, R.J. (1973). A possible mechanism for the inhibition of plant viruses by a peptide from Phytolacca americana. Virology 56, 390-393.

Palukaitis, P., and Zaitlin, M. (1984). A model to explain the "cross-protection" phenomenon shown by plant viruses and viroids. In Plant-Microbe Interactions. Molecular and Genetic Persepectives. Vol. 1. (Edited by T. Kosuge and E.W. Nester). pp. 420-429. Macmillan, New York and London.

Panopoulos, N.J., Faccioli, G., and Gold, A.H. (1972). Kinetics of carbohydrate metabolism in curly top virus infected tomato plants. Phytopathol. Mediterr. 12, 48-58.

Parent, J.G., and Asselin, A. (1984). Detection of pathogenesis-related (PR or b) and of other proteins in the intercellular fluid of hypersensitive plants infected with tobacco mosaic virus. Can. J. Bot. 62, 564-569.

Paterson, R., and Knight, C.A. (1975). Protein synthesis in tobacco protoplasts infected with tobacco mosaic virus. Virology 64, 10-22.

Paulsen, A.Q., and Sill, W.H. (1970). Absence of cross protection between maize dwarf virus strains A and B in grain sorghums. Plant Dis. Rep. 54, 627-629.

Pelcher, L.E., Walmsley, S.L., and Mackenzie, S.L. (1980). The effects of heterologous and homologous coat protein on alkaline disassembly of tobacco and tomato isolates of tobacco mosaic virus. Virology 105, 287-290.

Pelham, J. (1972). Strain-genotype interaction of tobacco mosaic virus in tomato. Ann. Appl. Biol. 71, 219-228.

Pennazio, S., D'Agostino, G., Appiano, A., and Redolfi, P. (1978). Ultrastructure and histochemistry of the resistant tissue surrounding lesions of tomato bushy stunt virus in Gomphrena globosa leaves. Physiol. Plant Pathol. 13, 165-171.

Pennazio, S., and Redolfi, P. (1980). Studies on the partly localized reaction of Gomphrena globosa leaves to potato virus X. Phytopathol. Mediterr. 19, 97-102.

Pennazio, S., and Sapetti, C. (1982). Electrolyte leakage in relation to viral and abiotic stresses inducing necrosis in cowpea leaves. Biol. Plant. 24, 218-255.

Perham, R.N., and Wilson, T.M.A. (1978). The characterization of intermediates formed during the disassembly of tobacco mosaic virus at alkaline pH. Virology 84, 293-302.

Person, C. (1959). Gene-for-gene relationships in host:parasite systems. Can. J. Bot. 37, 1101-1130.

Peterson, P.D., and McKinney, H.H. (1938). The influence of four mosaic diseases on the plastid pigments and chlorophyllase in tobacco leaves. Phytopathology 28, 329-342.

Piazolla, P., Tousignant, M.E., and Kaper, J.M. (1982). Cucumber mosaic virus-associated RNA 5. IX. The overtaking of viral RNA synthesis by CARNA 5 and dsCARNA 5 in tobacco. Virology 122, 147-157.

Pierpoint, W.S. (1983a). The major proteins in extracts of tobacco leaves that are responding hypersensitively to virus infection. Phytochemistry 22, 2691-2697.

Pierpoint, W.S. (1983b). Is there a phytointerferon? Trends Biochem. Sci. 8, 5-7.

Pierpoint, W.S., Robinson, N.P., and Leason, M.B. (1981). The pathogenesis-related proteins of tobacco: their induction by virus in intact plants and their induction by chemicals in detached leaves. Physiol. Plant Pathol. 19, 85-97.

Pitrat, M., and Lecoq, H. (1980). Inheritance of resistance to cucumber mosaic virus transmission by *Aphis gossypii* in *Cucumis melo*. Phytopathology 70, 958-961.

Platt, S.G., Hendriques, F., and Rand, L. (1979). Effects of virus infection on the chlorophyll content, photosynthetic rate and carbon metabolism of *Tolmiea menziesii*. Physiol. Plant Pathol. 15, 351-365.

Porter, C.A. (1959). Biochemistry of plant virus infections. Adv. Virus Res. 6, 75-93.

Pring, D.R. (1971). Viral and host RNA synthesis in BSMV-infected barley. Virology 44, 54-66.

Pritchard, D.W., and Ross, A.F. (1975). The relationship of ethylene to formation of tobacco mosaic virus lesions in hypersensitive responding tobacco leaves with and without induced resistance. Virology 64, 295-307.

Rackwitz, H.R., Rhode, W., and Sänger, H.L. (1981). DNA-dependent RNA polymerase II of plant origin transcribes viroid RNA into full-length copies. Nature 291, 297-301.

Rajagopal, R. (1977). Effect of tobacco mosaic virus infection on the endogenous levels of indoleacetic, phenylacetic and abscisic acids of tobacco leaves in various stages of development. Z. Pflanzenphysiol. 83, 403-409.

Rajagopalan, N., and Raju, P. (1972). The influence of infection by dolichos enation mosaic virus on nodulation and nitrogen fixation by field bean (*Dolichos lablab* L.). Phytopathol. Z. 73, 241-259.

Rajagopal, R. (1977). Effect of tobacco mosaic virus infection on the endogenous levels of indoleacetic, phenylacetic and abscisic acids of tobacco leaves in various stages of development. Z. Pflanzenphysiol. 83, 403-409.

Ralph, R.K., Bullivant, S., and Wojcik, S.J. (1971). Evidence for the intracellular site of double-stranded turnip yellow mosaic virus RNA. Virology 44, 473-479.

Randles, J.W. (1975). Association of two ribonucleic acid species with cadang-cadang disease of coconut palm. Phytopathology 65,

163-167.

Randles, J.W., and Coleman, D.F. (1970). Loss of ribosomes in *Nicotiana glutinosa* L. infected with lettuce necrotic yellows virus. Virology 41, 459-464.

Randles, J.W., and Coleman, D. (1972). Changes in polysomes in *Nicotiana glutinosa* L. leaves infected with lettuce necrotic yellows virus. Physiol. Plant Pathol. 2, 247-258.

Randles, J.W., and Francki, R.I.B. (1972). Infectious nucleocapsid particles of lettuce necrotic yellows virus with RNA-dependent RNA polymerase activity. Virology 50, 297-300.

Rast, A.T.B. (1975). Variability of tobacco mosaic virus in relation to control of tomato mosaic in glasshouse crops by resistance breeding and cross protection. Agric. Res. Rep. (Wageningen) 834, 76 pp.

Reanney, D.C. (1982). The evolution of RNA viruses. Annu. Rev. Microbiol. 36,, 47-73.

Reanney, D.C., and Ackermann, H.W. (1982). Comparative biology and evolution of bacteriophages. Adv. Virus Res. 27, 205-280.

Reddi. K.K. (1963). Studies on the formation of tobacco mosaic virus. II. Degradation of host ribonucleic acid following infection. Proc. Natl. Acad. Sci. USA. 50, 75-81.

Reddy, D.V.R., Rhodes, D.P., Lesnaw, J.A., MacLeod, R., Banerjee, A.K., and Black, L.M. (1977). *In vitro* transcription of wound tumour virus RNA by virion-associated RNA transcriptase. Virology 80, 356-361.

Redolfi, P. (1983). Occurrence of pathogenesis-related (b) and similar proteins in different plant species. Neth. J. Plant Pathol. 89, 245-254.

Reichman, M., Devash, Y., Suhadolnik, R.J., and Sela, I. (1983). Human leukocyte interferon and the antiviral factor (AVF) from virus-infected plants stimulate plant tissues to produce nucleotides with antiviral activity. Virology 128, 240-244.

Reid, M.S., and Matthews, R.E.F. (1966). On the origin of the mosaic induced by turnip yellow mosaic virus. Virology 28, 563-570.

Rezelman, G., Franssen, H.J., Goldbach, R.W., Ie, T.S., and Van Kammen, A. (1982). Limits to the independence of bottom component RNA of cowpea mosaic virus. J. Gen. Virol. 60, 335-342.

Rhodes, D.P., Reddy, D.V.R., MacLeod, R., Black, L.M., and Banerjee, A.K. (1977). *In vitro* synthesis of RNA containing 5' terminal structures ^{7}mG(5')ppp(5')Ap^{m} by purified wound tumour virus. Virology 76, 554-559.

Roberts, D.A., and Corbett, M.K. (1965). Reduced photosynthesis in tobacco plants infected with tobacco ringspot virus. Phytopathology 55, 370-371.

Roberts, P.L., and Wood. K.R. (1981a). Decrease in ribosome levels in tobacco infected with a chlorotic strain of cucumber mosaic virus. Physiol. Plant Pathol. 19, 99-111.

Roberts, P.L., and Wood, K.R. (1981b). Comparison of protein synthesis in tobacco systemically infected with a severe or a mild strain of cucumber mosaic virus. Phytopathol. Z. 102, 257-265.

Roberts, P.L., and Wood, K.R. (1982). Effects of a severe (P6) and a mild (W) strain of cucumber mosaic virus on tobacco leaf chlorophyll, starch and cell ultrastructure. Physiol. Plant Pathol. 21, 31–37.

Robinson, R.A. (1976). Plant Pathosystems. Springer–Verlag, Berlin. 184 pp.

Rodriguez, J.L., Garcia–Martinez, J.L., and Flores, R. (1978). The relationship between plant growth substance content and infection of Gynura aurantiaca DC by citrus exocortis viroid. Physiol. Plant Pathol. 13, 355–363.

Rohloff, H., and Lerch, B. (1977). Soluble leaf proteins in virus infected plants and acquired resistance. I. Investigations on Nicotiana tabacum cvs 'Xanthi-nc' and 'Samsun'. Phytopathol. Z. 89, 306–317.

Romaine, C.P., and Zaitlin, M. (1978). RNA dependent RNA polymerases in uninfected and tobacco mosaic virus–infected tobacco leaves: viral induced stimulation of a host polymerase activity. Virology 86, 241–253.

Roosien, J., Sarachu, A.N., Alblas, F., and Van Vloten–Doting, L. (1983). An alfalfa mosaic virus RNA–2 mutant which does not induce a hypersensitive reaction in cowpea plants, is multiplied to a high concentration in cowpea protoplasts. Plant Mol. Biol. 2, 85–88.

Ross, A.F. (1961a). Localized acquired resistance to plant virus infection in hypersensitive hosts. Virology 14, 329–339.

Ross, A.F. (1961b). Systemic resistance induced by localized virus infections in plants. Virology 14, 340–358.

Ross, A.F. (1966). Systemic effects of local lesion formation. In Viruses of Plants. (Edited by A.B.R. Beemster and J. Dijkstra). pp. 127–150. North Holland, Amsterdam.

Ross, A.F., and Israel, H.W. (1970). Use of heat treatments in the study of acquired resistance to tobacco mosaic virus in hypersensitive tobacco. Phytopathology 60, 755–770.

Ross, A.F., and Williamson, C.E. (1951). Physiologically active emanations from virus–infected plants. Phytopathology 41, 431–438.

Rottier, P.J.M., Rezelman, G., and Van Kammen, A. (1979). The inhibition of cowpea mosaic virus replication by actinomycin D. Virology 92, 299–309.

Russell, S.L., and Kimmins, W.C. (1971). Growth regulators and the effect of BYDV on barley (Hordeum vulgare L.). Ann. Bot. 35, 1037–1043.

Russo,, M., Martelli, G.P., and Franco, A. di (1981). The fine structure of local lesions of beet necrotic yellow vein virus in Chenopodium amaranticolor. Physiol. Plant Pathol. 19, 237–242.

Ruzicska, P., Gombos, Z., and Farkas, G.L. (1983). Modification of the fatty acid composition of phospholipids during the hypersensitive reaction in tobacco. Virology 128, 60–64.

Sakai, F., and Takebe, I. (1972). A non–coat protein synthesized in tobacco mesophyll protoplasts infected by tobacco mosaic virus. Mol. Gen. Genet. 118, 93–96.

Sakai, F., and Takebe, I. (1974). Protein synthesis in tobacco mesophyll protoplasts induced by tobacco mosaic virus infection. Virology 62, 426–433.

Saksena, K.N., and Mink, G.I. (1970). The effects of oxidized phenolic compounds on the infectivity of four "stable" viruses. Virology 40, 540–546.

Sandfaer, J. (1979). The influence of different strains of barley stripe mosaic virus on the frequency of triploids and aneuploids in barley. Phytopathol. Z. 95, 97–104.

Sänger, H.L. (1982). Biology, structure, functions and possible origin of viroids. In Encyclopedia of Plant Physiology, New Series, volume 14B. (Edited by B. Parthier and D. Boulter). pp. 369–454. Springer-Verlag, Berlin.

Sänger, H.L. (1984). Minimal infectious agents: the viroids. In The Microbe, 1984. Part 1, Viruses. (Edited by B.W.J. Mahy and J.R. Pattison). pp. 281–334. Cambridge University Press, Cambridge.

Sarkar, S., and Smitamana, P. (1981). A proteinless mutant of tobacco mosaic virus: evidence against the role of a viral coat protein for interference. Mol. Gen. Genet. 184, 158.

Scalla, R., Romaine, C.P., Asselin, A., Rigaud, J., and Zaitlin, M. (1978). An in vivo study of a nonstructural polypeptide synthesized upon TMV infection and its identification with a polypeptide synthesized in vitro from TMV RNA. Virology 91, 182–193.

Schnölzer, M, Haas, B., Ramm, K., Hofmann, H., and Sänger, H.L. (1985). Correlation between structure and pathogenicity of potato spindle tuber viroid (PSTV). EMBO J. 4, 2181–2190.

Schuch, W. (1974). Effects of two protein defective strains of TMV on the stability of chloroplast ribosomal RNA. Virology 60, 592–594.

Schumacher, J., Sänger, H.L., and Riesner, D. (1983). Subcellular localization of viroids in highly purified nuclei from tomato leaf tissue. EMBO J. 2, 1549–1555.

Schuster, G., and Wetzler, C. (1982). On virus-induced inhibitors from locally TMV-infected plants of Nicotiana glutinosa L. Phytopathol. Z. 104, 46–59.

Scott, S.W. (1982). Tests for resistance to white clover mosaic virus in red and white clover. Ann. Appl. Biol. 100, 393–398.

Sela, I. (1981a). Plant-virus interactions related to resistance and localization of viral infections. Adv. Virus Res. 26, 201–238.

Sela, I. (1981b). Antiviral factors from virus-infected plants. Trends Biochem. Sci. 6, 31–33.

Sela, I., and Harpaz, I. (1977). Further studies on the biology of an antiviral factor (AVF) from virus infected plants and its association with the N-gene of Nicotiana species. J. Gen. Virol. 35, 107–116.

Sela, I., Harpaz, I., and Birk, Y. (1964). Separation of a highly active antiviral factor from virus-infected plants. Virology 22, 446–451.

Sela, I., Hauschner, A., and Mozes, R. (1978). The mechanisms of stimulation of the antiviral factor (AVF) in Nicotiana leaves. The involvement of phosphorylation and the role of the N-gene. Virology 89, 1–6.

Selman, I.W., Brierley, M.R., Pegg, G.F., and Hill, T.A. (1961). Changes in the free amino acids and amides in tomato plants inoculated with tomato spotted wilt virus. Ann. Appl. Biol. 49, 601–615.

Semancik, J.S., and Geelen, J.L.M.C. (1975). Detection of DNA complementary to pathogenic viroid RNA in exocortis disease. Nature 256, 753–756.

Sequiera, L. (1984). Cross protection and induced resistance: their potential for plant disease control. Trends Biotechnol. 2, 25–29.

Shalla, T.A. (1964). Assembly and aggregation of tobacco mosaic virus in tomato leaflets. J. Cell Biol. 21, 253–264.

Shalla, T.A. and Peterson, L.J. (1978). Studies on the mechanism of viral cross protection. Phytopathology 68, 1681–1683.

Shalla, T.A., Petersen, L.J., and Zaitlin, M. (1982). Restricted movement of a temperature-sensitive virus in tobacco leaves is associated with a reduction in numbers of plasmodesmata. J. Gen. Virol. 60, 355–358.

Shaw, J.G. (1969). In vivo removal of protein from tobacco mosaic virus after inoculation of tobacco leaves. II. Some characteristics of the reaction. Virology 37, 109–116.

Shaw, J.G. (1970). Uncoating of tobacco mosaic virus RNA after inoculation of tobacco leaves. Virology 42, 41–48.

Shaw, J.G. (1985). Early events in plant virus infection. In Molecular Plant Virology Vol. II. (Edited by J. W. Davies). pp. 1–21. CRC Press, Boca Raton.

Shaw, J.G., Plaskitt, K.A., and Wilson, T.M.A. (1986). Evidence that tobacco mosaic virus particles disassembly co-translationally in vivo. Virology 148, 326–336.

Sheen, S.J., and Diachun, S. (1978). Peroxidases of red clovers resistant and susceptible to bean yellow mosaic virus. Acta Phytopathol. Acad. Sci. Hung. 13, 21–28.

Sheen, S.J., and Lowe, R.H. (1979). Proteins and related nitrogenous compounds in virus-infected tobacco plants. Can. J. Plant Sci. 59, 1099–1107.

Shepardson, S., Esau, K., and McCrum, R. (1980). Ultrastructure of potato leaf phloem infected with potato leafroll virus. Virology 105, 379–392.

Shepherd, R.J., Richins, R., and Shalla, T.A. (1980). Isolation and properties of the inclusion bodies of cauliflower mosaic virus. Virology 102, 389–400.

Sherwood, J.L., and Fulton, R.W. (1982). The specific involvement of coat protein in tobacco mosaic virus cross protection. Virology 119, 150–158.

Sherwood, J.L., and Fulton, R.W. (1983). Competition for infection sites and multiplication of the competing strain in plant viral interference. Phytopathology 73, 1363–1365.

Shockey, M.W., Gardener, C.O., Melcher, U., and Essenberg, R.C. (1980). Polypeptides associated with inclusion bodies from leaves of turnip infected with cauliflower mosaic virus. Virology 105, 575–581.

Siegel, A. (1985). Plant-virus-based vectors for gene transfer may be of considerable use despite a presumed high error frequency during RNA synthesis. Plant Mol. Biol. 4, 327–329.

Siegel, A., Hari, V., and Kolacz, K. (1978). The effect of tobacco mosaic virus infection on host and virus-specific protein synthesis in protoplasts. Virology 85, 494–503.

Siegel, A., Hari, V., Montgomery, I., and Kolacz, K. (1976). A messenger RNA for capsid protein isolated from tobacco mosaic virus-infected tissue. Virology 73, 363-371.
Sill, W.H., and Walker, J.C. (1952). A virus inhibitor in cucumber in relation to mosaic resistance. Phytopathology 42, 349-352.
Simons, T.J., and Ross, A.F. (1971a). Metabolic changes associated with systemic induced resistance to tobacco mosaic virus in Samsun NN tobacco. Phytopathology 61, 293-300.
Simons, T.J., and Ross, A.F. (1971b). Changes in phenol metabolism associated with induced systemic resistance to tobacco mosaic virus in Samsun NN tobacco. Phytopathology 61, 1261-1265.
Simpson, R.S., Chakravorty, A.K., and Scott, K.J. (1979). Selective hydrolysis of barley leaf polysomal messenger RNA during the early stages of powdery mildew infection. Physiol. Plant Pathol. 14, 245-258.
Singer, B., and Condit, C. (1974). Protein synthesis in virus-infected plants. III. Effects of tobacco mosaic virus mutants on protein synthesis in Nicotiana tabacum. Virology 57, 42-48.
Singh, B.P., and Mohan, J. (1982). Influence of cucumis mosaic virus-1 on free amino acids of Cucumis sativus. Acta. Bot. Indica 19, 308-310.
Singh, R., and Singh, H.C. (1982). Nitrate reductase activity in cowpea mosaic virus infected cowpea plant parts. Acta Microbiol. Pol. 31, 95-98.
Singh, S.J. (1983). Changes in enzymatic activity of pumpkin plant infected with watermelon mosaic virus. J. Turkish Phytopathol. 12, 33-38.
Sisco, P.H., Garcia-Arenal, F., Zaitlin, M., Earle, E.D., and Gracen, V.E. (1984). LBN, a male-sterile cytoplasm of maize, contains two double-stranded RNAs. Plant Sci. Lett. 34, 127-134.
Skaria, M., Lister, R.M., Forster, J.E., and Shaner, G. (1985). Virus content as an index of symptomatic resistance to barley yellow dwarf virus in cereals. Phytopathology 75, 212-216.
Smith, P.R., and Neales, T.F. (1977a). Analysis of the effects of virus infection on the photosynthetic properties of peach leaves. Aust. J. Plant Physiol. 4, 723-732.
Smith, P.R., and Neales, T.F. (1977b). The growth of young peach trees following infection by the viruses of peach rosette and decline disease. Aust. J. Agric. Res. 28, 441-444.
Smith, P.R., and Sward, R.J. (1982). Crop loss assessment studies on the effects of barley yellow dwarf virus in wheat in Victoria. Aust. J. Agric. Res. 33, 179-185.
Smith, S.H., McCall, S.R., and Harris, J.H. (1968). Alterations in the auxin levels of resistant and susceptible hosts induced by the curly top virus. Phytopathology 58, 575-577.
Solymosy, F., and Farkas, G.L. (1962). Simultaneous activation of pentose-phosphate shunt enzymes in a virus-infected local lesion host plant. Nature 195, 835.
Solymosy, F., and Farkas, G.L. (1963). Metabolic characteristics at the enzymatic level of tobacco tissues exhibiting localized acquired resistance to viral infection. Virology 21, 210-221.

Solymosy, F., and Kiss, T. (1985). Viroids and snRNAs. In Subviral Pathogens of Plants and Animals: Viroids and Prions. (Edited by K. Maramorosch and J.J. McKelvey). pp. 183-199. Academic Press, Orlando.

Solymosy, F., Szirmai, J., Beczner, L., and Farkas, G.L. (1967). Changes in peroxidase isoenzyme patterns induced by virus infection. Virology 32, 117-121.

Sprague, G.F., and McKinney, H.H. (1966). Abberant ratio: an anomaly in maize associated with virus infection. Genetics 54, 1287-1296.

Sprague, G.F., and McKinney, H.H. (1971). Further evidence on the genetic behaviour of AR in maize. Genetics 67, 533-542.

Stein, A., and Loebenstein, G. (1976). Peroxidase activity in tobacco plants with polyanion-induced interference to tobacco mosaic virus. Phytopathology 66, 1192-1194.

Stein, A., Loebenstein, G., and Spiegel, S. (1979). Further studies of induced interference by a synthetic polyanion of infection by tobacco mosaic virus. Physiol. Plant Pathol. 15, 241-255.

Stirpe, F., Williams, D.G., Onyon, L.J., Legg, R.F., and Stevens, W.A. (1981). Dianthins, ribosome-damaging proteins with antiviral properties from *Dianthus caryophyllus* L. (carnation). Biochem. J. 195, 399-405.

Sulzinski, M.A., Gabard, K.A., Palukaitis, P., and Zaitlin, M. (1985). Replication of tobacco mosaic virus. VIII. Characterization of a third subgenomic TMV RNA. Virology 145, 132-140.

Sulzinski, M.A., and Zaitlin, M. (1982). Tobacco mosaic virus replication in resistant and susceptible plants: in some resistant species virus is confined to a small number of initially infected cells. Virology 121, 12-19.

Sunderland, D.W., and Merrett, M.J. (1964). Adenosine diphosphate and adenosine triphosphate concentrations in leaves showing necrotic local virus lesions. Virology 23, 274-276.

Sunderland, D.W., and Merrett, M.J. (1965). The respiration of leaves showing necrotic local lesions following infection by tobacco mosaic virus. Ann. Appl. Biol. 56, 477-484.

Symons, R.H., Haseloff, J., Visvader, J.E., Keese, P., Murphy, P.J., Gill, D.S., Gordon, K.H.J., and Bruening, G. (1985). On the mechanism of replication of viroids, virusoids and satellite RNAs. In Subviral Pathogens of Plants and Animals: Viroids and Prions. (Edited by K. Maramorosch and J.J. McKelvey). pp. 235-263. Academic Press, Orlando.

Sziraki, I., and Balazs, E. (1975). The effect of infection by TMV on cytokinin level of tobacco plants, and cytokinins in TMV-RNA. In Current Topics in Plant Pathology. (Edited by Z. Kiraly). pp. 345-352. Akademiai Kiado, Budapest.

Sziraki, I., and Balazs, E. (1979). Cytokinin activity in the RNA of tobacco mosaic virus. Virology 92, 578-582.

Sziraki, I., Balazs, E., and Kiraly, Z. (1980). Role of different stresses in inducing systemic acquired resistance to TMV and increasing cytokinin level in tobacco. Physiol. Plant Pathol. 16, 277-284.

Sziraki, I., and Gaborjanyi, R. (1974). Effect of systemic infection by TMV on cytokinin level of tobacco leaves and stems. Acta Phytopathol. Acad. Sci. Hung. 9, 195-199.

Takahashi, T. (1973). Studies on viral pathogenesis in plant hosts. IV. Comparison of early processes of tobacco mosaic virus infection in the leaves of 'Samsun NN' and 'Samsun' tobacco plants. Phytopathol. Z. 77, 157-168.

Takanami, Y. (1981). A striking change in symptoms on cucumber mosaic virus-infected tobacco plants induced by a satellite RNA. Virology 109, 120-126.

Takebe, I., and Otsuki, Y. (1969). Infection of tobacco mesophyll protoplasts by tobacco mosaic virus. Proc. Natl. Acad. Sci. USA. 64, 843-848.

Takusari, H., and Takahashi, T. (1979). Studies on viral pathogenesis in plant hosts: IX. Effect of citrinin on the formation of necrotic lesions and virus localization in the leaves of 'Samsun NN' tobacco plants after tobacco mosaic virus infection. Phytopathol. Z. 96, 324-329.

Taliansky, M.E., Malyshenko, S.I., Pshennikova, E.S., and Atabekov, J.G. (1982). Plant virus-specific transport function. II. A factor controlling virus host range. Virology 122, 327-331.

Tanguy, J., and Martin, C. (1972). Phenolic compounds and the hypersensitivity reaction in Nicotiana tabacum infected with tobacco mosaic virus. Phytochemistry 11, 19-28.

Taniguchi, T. (1963). Similarity in the accumulation of tobacco mosaic virus in systemic and local necrotic infection. Virology 19, 237-238.

Tarn, T.R., and Adams, J.B. (1982). Aphid probing and feeding, electronic monitoring and plant breeding, in Pathogens, Vectors and Plant Diseases, Approaches to Control. (Edited by K.F. Harris and K. Maramorosch). pp. 221-246. Academic Press, New York.

Tas, P.W.L., and Peters, D. (1977). The occurrence of a soluble protein (E_1) in cucumber cotyledons infected with plant viruses. Neth. J. Plant Pathol. 83, 5-12.

Tavantzis, S.M. (1984). The use of terms for responses of plants to viruses: a reply to recent proposals. Phytopathology 74, 379-380.

Tavantzis, S.M., Smith, S.H., and Witham, F.H. (1979). The influence of kinetin on tobacco ringspot virus infectivity and the effect of virus infection on the cytokinin activity in intact leaves of Nicotiana glutinosa L. Physiol. Plant Pathol. 14, 227-233.

Thomas, P., and Fulton, R.W. (1968). Correlation of the ectodesmata number with non-specific resistance to initial virus infection. Virology 34, 459-469.

Tine, P., and Chang, X.H. (1983). Control of two seed-borne virus diseases in China by the use of protective inoculum. Seed. Sci. Technol. 11, 969-972.

Tomenius, K., and Oxelfelt, O. (1982). Ultrastructure of pea leaf cells infected with three strains of red clover mottle virus. J. Gen. Virol. 61, 143-147.

Tomlinson, J.A. (1982). Chemotherapy of plant virus disease. In Pathogens, Vectors and Plant Diseases. (Edited by K.F. Harris and K. Maramorosch). pp. 23-44. Academic Press, New York.

Tomlinson, J.A., Carter, A.L., Dale, W.T., and Simpson, C.J. (1970).

Weed plants as sources of cucumber mosaic virus. Ann. Appl. Biol. 66, 11-16.

Tomlinson, J.A., Faithfull, E.M., and Ward, C.M. (1976). Chemical suppression of the symptoms of two virus diseases. Ann. Appl. Biol. 84, 31-41.

Tomlinson, J.A., and Webb, M.J.W. (1978). Ultrastructural changes in chloroplasts of lettuce infected with beet western yellows virus. Physiol. Plant Pathol. 12, 13-18.

Toriyama, S., and Peters, D. (1980). In vitro synthesis of RNA by dissociated lettuce necrotic yellows virus particles. J. Gen. Virol. 50, 125-134.

Trewavas, A. (1981). How do plant growth substances work? Plant Cell Environ. 4, 203-228.

Tsugita, A. (1962). The proteins of mutants of TMV: composition and structure of chemically evoked mutants of TMV RNA. J. Mol. Biol. 5, 284-292.

Tsugita, A., and Fraenkel-Conrat, H. (1960). The amino acid composition and C-terminal sequence of a chemically evoked mutant of TMV. Proc. Natl. Acad. Sci. USA. 46, 636-642.

Tu, J.C. (1977). Cyclic AMP in clover and its possible role in clover yellow mosaic virus-infected tissue. Physiol. Plant Pathol. 10, 117-123.

Tu, J.C., Ford, R.E., and Gray, C.R. (1970a). Some factors affecting the nodulation and nodule efficiency of soybeans infected by soybean mosaic virus. Phytopathology 60, 1653-1656.

Tu, J.C., Ford, R.E., and Krass, C.J. (1968). Effect of maize dwarf mosaic virus infection on respiration and photosynthesis of corn. Phytopathology 58, 282-284.

Tu, J.C., Ford, R.E., and Quiniones, S.S. (1970b). Effect of soybean mosaic virus and (or) bean pod mottle virus infection on soybean nodulation. Phytopathology 60, 518-523.

Turner, M.S., and Dawson, W.O. (1984). Specificity of the actinomycin-D sensitive function of some plant viruses. Intervirology 21, 224-228.

Tuzun, S., and Kuc, J. (1985). A modified technique for inducing systemic resistance to blue mold and increasing growth in tobacco. Phytopathology 75, 1127-1129.

Ushiyama, R., and Matthews, R.E.F. (1970). The significance of chloroplast abnormalities associated with infection by turnip yellow mosaic virus. Virology 42, 293-303.

Valverde, R.A., and Fulton, J.P. (1982). Characterization and variability of strains of southern bean mosaic virus. Phytopathology 72, 1265-1268.

Valverde, R.A., Moreno, R., and Gamez, R. (1982). Yield reduction in cowpea (Vigna unguiculata L. Walp.) infected with cowpea mosaic virus in Costa Rica. Turrialba 32, 89-90.

Van, K.T.T., Tonbart, P., Cousson, A., Darvill, A.G., Gollin, D.J., Chelf, P., and Albersheim, P. (1985). Manipulation of the morphogenetic pathways of tobacco explants by oligosaccharins. Nature 314, 615-617.

Van der Meer, J., Dorssers, L., Van Kammen, A., and Zabel, P. (1984). The RNA-dependent RNA polymerase of cowpea is not involved in cowpea mosaic virus replication: immunological evidence. Virology 132, 413-425.

Vanderplank, J.E. (1984). Disease Resistance in Plants (2nd edition). Academic Press, Orlando, Florida. 194 pp.

Van Kammen, A., and Brouwer, D. (1964). Increase of polyphenoloxidase activity by a local virus infection in uninoculated parts of the leaf. Virology 22, 9-14.

Van Loon, L.C. (1975). Polyacrylamide disk electrophoresis of the soluble leaf proteins from Nicotiana tabacum var. Samsun and Samsun NN. Physiol. Plant Pathol. 6, 289-300.

Van Loon, L.C. (1976). Specific soluble leaf proteins in virus-infected plants are not normal constituents. J. Gen. Virol. 30, 375-379.

Van Loon, L.C. (1977). Induction by 2-chloroethylphosphonic acid of viral-like lesions, associated proteins and systemic resistance in tobacco. Virology 80, 417-420.

Van Loon, L.C. (1982). Regulation of changes in proteins and enzymes associated with active defence against virus infection. In Active Defense Mechanisms in Plants. (Edited by R.K.S. Wood). pp. 247-273. Plenum Press, New York.

Van Loon, L.C. (1983a). Mechanisms of resistance in virus-infected plants. In The Dynamics of Host Defence. (Edited by J.A. Bailey and B.J. Deverall). pp. 123-190. Academic Press, Australia.

Van Loon, L.C. (1983b). The induction of pathogenesis-related proteins by pathogens and specific chemicals. Neth. J. Plant Pathol. 89, 265-273.

Van Loon, L.C. (1985). Pathogenesis-related proteins. Plant Mol. Biol. 4, 111-116.

Van Loon, L.C., and Antoniw, J.F. (1982). Comparison of the effects of salicylic acid and ethephon with virus-induced hypersensitivity and acquired resistance in tobacco. Neth. J. Plant Pathol. 88, 237-256.

Van Loon, L.C., and Berbee, A.T. (1978). Endogenous levels of indoleacetic acid in leaves of tobacco reacting hypersensitively to TMV. Z. Pflanzenphysiol. 89, 373-375.

Van Loon, L.C., and Djikstra, J. (1976). Virus-specific expression of systemic acquired resistance in tobacco mosaic virus- and tobacco necrosis virus-infected 'Samsun NN' and 'Samsun' tobacco. Neth. J. Plant Pathol. 82, 231-237.

Van Loon, L.C., and Geelen, J.L.M.C. (1971). The relation of polyphenoloxidase and peroxidase to symptom expression in tobacco var. 'Samsun NN' after infection with tobacco mosaic virus. Acta Phytopathol. Acad. Sci. Hung. 6, 9-20.

Van Loon, L.C., and Van Kammen, A. (1970). Polyacrylamide disc electrophoresis of the soluble leaf proteins from N. tabacum var. Samsun and Samsun NN. II. changes in protein constitution after infection with TMV. Virology 40, 199-211.

Van Steveninck, R.F.M. (1959). Factors affecting the abscission of reproductive organs in yellow lupins (Lupinus luteus L.). III. endogenous growth substances in virus-infected and healthy plants and their effect on abscission. J. Exp. Bot. 10, 367-376.

Van Telgen, H.J., Goldbach, R.W., and Van Loon, L.C. (1985a). The

126,000 molecular weight protein of tobacco mosaic virus is associated with host chromatin in mosaic-diseased tobacco plants. Virology 143, 612-616.

Van Telgen, H.J., Van der Zaal, E.J., and Van Loon, L.C. (1985b). Some characteristics of the association of the 116 kD protein with host chromatin in tobacco leaves infected with tobacco mosaic virus. Physiol. Plant Pathol. 26, 99-110.

Van Telgen, H.J., Van der Zaal, E.J., and Van Loon, L.C. (1985c). Evidence for an association between viral coat protein and host chromatin in mosaic-diseased tobacco leaves. Physiol. Plant Pathol. 26, 83-98.

Vegetti, G., Conti, G.G., and Pesci, P. (1975). Changes in phenylalanine ammonia lyase, peroxidase and polyphenoloxidase during development of local necrotic lesions in pinto bean leaves infected with alfalfa mosaic virus. Phytopathol. Z. 84, 153-171.

Venekamp, J., and Beemster, A.B.R. (1980). Mature plant resistance of potato against some virus diseases. I. Concurrence of development of mature plant resistance against potato virus X and decrease of ribosome and RNA content. Neth. J. Plant Pathol. 86, 1-10.

Venekamp, J.H., Schepers, A., and Bus, C.B. (1980). Mature plant resistance of potato against some virus diseases. III. Mature plant resistance against potato virus Y^N indicated by decrease in ribosome content in ageing potato plants under field conditions. Neth. J. Plant Pathol. 86, 301-309.

Verduin, B.J.M., Prescott, B., and Thomas, G.J. (1984). RNA-protein interactions and secondary structures of cowpea chlorotic mottle virus for *in vitro* assembly. Biochemistry 23, 4301-4308.

Verma, H.N., and Awasthi, L.P. (1980). Occurrence of a highly antiviral agent in plants treated with *Boerhaavia diffusa* inhibitor. Can. J. Bot. 58, 2141-2144.

Verma, H.N., and Baranwal, V.K. (1983). Antiviral activity and the physical properties of the leaf extract of *Chenopodium ambrooides* L. Proc. Indian Acad. Sci. 92, 461-465.

Verma, H.N., Chowdhury, B., and Rastogi, P. (1984). Antiviral activity in leaf extracts of different *Clerodendron* species. Z. Pflanzenkr. Pflanzenschutz 91, 34-41.

Verma, H.N., and Dwivedi, S.D. (1984). Properties of a virus inhibiting agent, isolated from plants which have been treated with leaf extracts from *Bougainvillea spectabilis*. Physiol. Plant Pathol. 25, 93-101.

Visvader, J.E., and Symons, R.H. (1985). Eleven new sequence variants of citrus exocortis viroid and the correlation of sequence with pathogenicity. Nucleic Acids Res. 13, 2906-2920.

Volovitch, M., Modjtahedi, N., and Yot, P. (1984). RNA-dependent DNA polymerase activity in cauliflower mosaic virus-infected plant leaves. EMBO J. 3, 309-314.

Wade, B.L., and Zaumeyer, W.J. (1940). Genetic studies of resistance to alfalfa mosaic virus and of stringiness in *Phaseolus vulgaris*. J. Am. Soc. Agron. 32, 127.

Wagih, E.E., and Coutts, R.H.A. (1981). Similarities in the soluble protein profiles of leaf tissue following either a hypersensitive reaction to virus infection or plasmolysis. Plant Sci. Lett. 21,

61-69.
Wagih, E.E., and Coutts, R.H.A. (1982a). Peroxidase, polyphenoloxidase and ribonuclease in tobacco necrosis virus infected or mannitol osmotically stressed cowpea and cucumber tissue. I. Quantitative alterations. Phytopathol. Z. 104, 1-12.
Wagih, E.E., and Coutts, R.H.A. (1982b). Peroxidase, polyphenoloxidase and ribonuclease in tobacco necrosis virus infected or mannitol osmotically stressed cowpea and cucumber tissue. II. Qualitative alterations. Phytopathol. Z. 104, 124-137.
Walkey, D.G.A. (1980). Production of virus free plants. Acta Hortic. 88, 23-31.
Walkey, D.G.A., Creed, C., Delanay, H., and Whitwell, J.D. (1982). Studies on the reinfection and yield of virus-tested and commercial stocks of rhubarb cv. Timperley Early. Plant Pathol. 31, 253-261.
Wasuwat, S.L., and Walker, J.C. (1961). Relative concentration of cucumber mosaic virus in a resistant and a susceptible cucumber variety. Phytopathology 51, 614-615.
Waterworth, H.E., Kaper, J.M., and Tousignant, M.E. (1979). CARNA 5, the small cucumber mosaic virus-dependent replicating RNA, regulates disease expression. Science 204, 845-847.
Weintraub, M., Ragetli, H.W.J., and Dwurazna, M.M. (1964). Studies on the metabolism of leaves with localized virus infections. Mitochondrial activity in TMV-infected *Nicotiana glutinosa*. Can. J. Bot. 42, 541-545.
Weintraub, M., Ragetli, H.W.J., and Lo, E. (1972). Mitochondrial content and respiration in leaves with localized virus infections. Virology 50, 841-850.
Weststeijn, E.A. (1976). Peroxidase activity in leaves of *Nicotiana tabacum* var. *Xanthi* nc. before and after infection with tobacco mosaic virus. Physiol. Plant Pathol. 8, 63-71.
Weststeijn, E.M. (1978). Permeability changes in the hypersensitive reaction of *Nicotiana tabacum* cv. Xanthi-nc after infection with tobacco mosaic virus. Physiol. Plant. Pathol. 13, 253-258.
Whenham, R.J., and Fraser, R.S.S. (1980). Stimulation by abscisic acid of RNA synthesis in discs from healthy and tobacco mosaic virus-infected tobacco leaves. Planta 150, 349-353.
Whenham, R.J., and Fraser, R.S.S. (1981). Effect of systemic and local-lesion-forming strains of tobacco mosaic virus on abscisic acid concentration in tobacco leaves: consequences for the control of leaf growth. Physiol. Plant Pathol. 18, 267-278.
Whenham, R.J., and Fraser, R.S.S. (1982). Does tobacco mosaic virus RNA contain cytokinins? Virology 118, 263-266.
Whenham, R.J., and Fraser, R.S.S. (1985). Abscisic acid metabolism in tobacco mosaic virus infected plants. Rep. Natl. Veg. Res. Stn for 1984, 23-24.
Whenham, R.J., Fraser, R.S.S., Brown, L., and Payne, J.A. (1986). Tobacco mosaic virus-induced increase in abscisic acid concentration in tobacco leaves: cellular location in light- and dark-green areas, and relationship to symptom development. Planta, in press.
Whenham, R.J., Fraser, R.S.S., and Snow, A. (1985). Tobacco mosaic

virus-induced increase in abscisic acid concentration in tobacco leaves: intracellular location and relationship to symptom severity and to extent of virus multiplication. Physiol. Plant Pathol. 26, 379-387.

White, J., and Brakke, M. (1982). Chloroplast RNA and proteins decrease as wheat streak and barley stripe mosaic viruses multiply in expanding, systemically infected leaves. Phytopathology 72, 939.

White, J.L., and Brakke, M.K. (1983). Protein changes in wheat infected with wheat streak mosaic virus and barley infected with barley stripe mosaic virus. Physiol. Plant Pathol. 22, 87-100.

White, J.L., and Dawson, W.O. (1978). Characterization of RNA-dependent RNA polymerases in uninfected and cowpea chlorotic mottle virus-infected cowpea leaves: selective removal of host RNA polymerase from membranes containing CCMV RNA replicase. Virology 88, 33-43.

White, R.F. (1979). Acetylsalicylic acid (aspirin) induces resistance to tobacco mosaic virus in tobacco. Virology 99, 410-412.

Wilkins, P.W., and Catherall, P.L. (1974). The effect of some isolates of ryegrass mosaic virus on different genotypes of *Lolium multiflorum*. Ann. Appl. Biol. 76, 209-216.

Williams, B.R.G., and Kerr, I.M. (1978). Inhibition of protein synthesis by 2'-5' linked adenine oligonucleotides in intact cells. Nature 276, 88-90.

Wilson, T.M.A. (1984a). Cotranslational disassembly of tobacco mosaic virus *in vivo*. Virology 137, 255-265.

Wilson, T.M.A. (1984b). Cotranslational disassembly increases the efficiency of expression of TMV RNA in wheat germ cell free extracts. Virology 138, 353-356.

Wilson, T.M.A. (1985). Nucleocapsid disassembly and early gene expression by positive-strand RNA viruses. J. Gen. Virol. 66, 1201-1207.

Wilson, T.M.A., and Glover, J.E. (1983). The origin of multiple polypeptides of molecular weight below 110,000 encoded by tobacco mosaic virus RNA in the messenger-dependent rabbit reticulocyte lysate. Biochem. Biophys Acta 739, 35-41.

Wilson, T.M.A., and Shaw, J.G. (1985). Does TMV uncoat co-translationally *in vivo*? Trends Biochem. Sci. 10, 57-60.

Wilson, T.M.A., and Watkins, P.A.C. (1985). Cotranslational disassembly of a cowpea strain (Cc) of TMV: evidence that viral RNA-protein interactions at the assembly origin block ribosome translocation *in vitro*. Virology 145, 346-349.

Wilson, T.M.A., and Watkins, P.A.C. (1986). Influence of exogenous viral coat protein on the cotranslational disassembly of tobacco mosaic virus (TMV) particles *in vitro*. Virology 149, 132-135.

Windham, M.T., and Ross, J.P. (1985). Phenotypic response of six soybean cultivars to bean pod mottle virus infection. Phytopathology 75, 305-309.

Wongkaew, S., and Peterson, J.F. (1983). Effect of peanut mottle virus infection on peanut nodulation and nodule function. Phytopathology 73, 377.

Wood, K.R. (1971). Peroxidase isoenzymes in leaves of cucumber

(Cucumis sativus L.) cultivars systemically infected with the W strain of cucumber mosaic virus. Physiol. Plant Path. 1, 133-139.

Wood, K.R., and Barbara, D.J. (1971). Virus multiplication and peroxidase activity in leaves of cucumber (Cucumis sativus L.) cultivars systemically infected with the W strain of cucumber mosaic virus. Physiol. Plant Pathol. 1, 73-81.

Woolston, C.J., Covey, S.N., Penswick, J.R., and Davies, J.W. (1983). Aphid transmission and a polypeptide are specified by a defined region of the cauliflower mosaic virus genome. Gene 23, 15-23.

Wu, J.H. (1964). Release of inhibited tobacco mosaic virus infection by ultraviolet irradiation as a function of time and temperature after inoculation. Virology 24, 441-445.

Wu, J.H., and Dimitman, J.E. (1970). Leaf structure and callose formation as determinants of tobacco mosaic virus movement in bean leaves as revealed by UV irradiation studies. Virology 40, 820-827.

Wu, J.H., Hudson, W., and Wildman, S.G. (1962). Quantitative analysis of interference produced by related strains of TMV compared with noninfectious TMV components. Phytopathology 52, 1264-1266.

Wunner, W.H. (1982). Is the acetylcholine receptor a rabies-specific receptor? Trends Neurosci. 5, 413-415.

Wyatt, S.D., and Kuhn, C.W. (1979). Replication and properties of cowpea chlorotic mottle virus in resistant cowpeas. Phytopathology 69, 125-129.

Wyatt, S.D., and Kuhn, C.W. (1980). Derivation of a new strain of cowpea chlorotic mottle virus from resistant cowpeas. J. Gen. Virol. 49, 289-296.

Wyatt, S.D., and Wilkinson, T.C. (1984). Increase and spread of cowpea chlorotic mottle virus in resistant and fully susceptible cowpeas. Physiol. Plant Pathol. 24, 339-345.

Wyen, N.V., Udvary, J., Erdei, S., and Farkas, G.L. (1972). The level of a relatively purine sensitive ribonuclease increases in virus-infected hypersensitive or mechanically injured tobacco leaves. Virology 48, 337-341.

Yamaguchi, A., and Hirai, T. (1959). The effect of local infection with tobacco mosaic virus on respiration in leaves of Nicotiana glutinosa. Phytopathology 49, 447-449.

Yarwood, C.E. (1959). Virus suceptibility increased by soaking bean leaves in water. Plant Dis. Rep. 43, 841-844.

Yarwood, C.E. (1960). Localized acquired resitance to tobacco mosaic virus. Phytopathology 50, 741-744.

Zadoks, J.C., and Schein, R.D. (1979). Epidemiology and Plant Disease Management. Oxford University Press, New York and Oxford. 426 pp.

Zaitlin, M. (1976). Viral cross protection: more understanding is needed. Phytopathology 66, 382-383.

Zaitlin, M. (1979). How viruses and viroids induce disease. In Plant Disease, an Advanced Treatise. (Edited by J.G.Horsfall and E.B.Cowling). Vol. IV. pp. 257-271. Academic Press, New York.

Zaitlin, M., and Hariharasubramanian, V. (1972). A gel electrophoretic analysis of proteins from plants infected with tobacco mosaic virus and potato spindle tuber viroid. Virology 47, 296-305.

Zaitlin, M., Niblett, C.L., Dickinson, E., and Goldberg, R.B. (1980). Tomato DNA contains no detectable regions complementary to potato spindle tuber viroid as assayed by solution and filter hybridization. Virology 104, 1-9.

Zeevaart, J.A.D., and Boyer, G.L. (1982). Metabolism of abscisic acid in Xanthium strumarium and Ricinus communis. In Plant Growth Substances 1982. (Edited by P.F. Wareing). pp. 335-342. Academic Press, London.

Ziemiecki, A., and Wood, K.R. (1975). Changes in the soluble protein constitution of cucumber cotyledons following infection with two strains of cucumber mosaic virus. Physiol. Plant Pathol. 7, 79-89.

Ziemiecki, A., and Wood, K.R. (1976). Proteins synthesized by cucumber cotyledons infected with two strains of cucumber mosaic virus. J. Gen. Virol. 31, 373-381.

Zimmern, D. (1982). Do viroids and RNA viruses derive from a system that exchanges genetic information between eukaryotic cells? Trends Biochem. Sci. 7, 205-207.

Index